U0070282

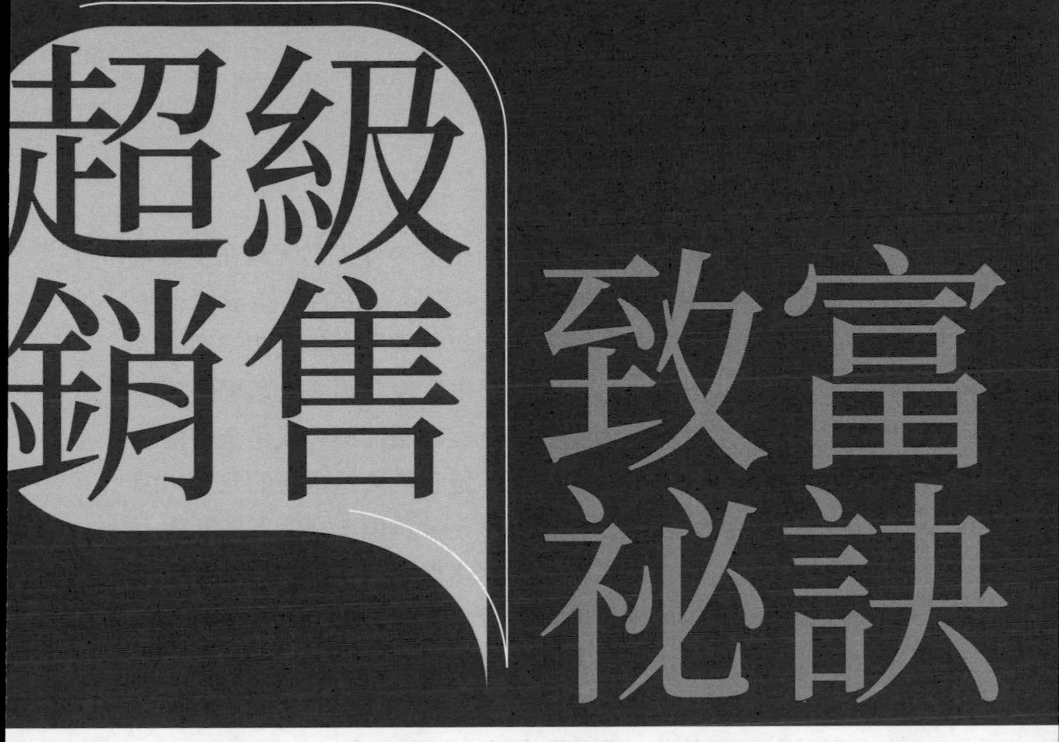

SECRET

of getting rich
through super sales

徐錦坤——編著

作者序一

吾人從事飲水機、淨水機事業已有 30 年左右，現在是井易飲水機租用加盟事業團隊負責人。基於現在的時空環境變化大，有眾多的社會人士，或眾多的年輕人和中年人想轉換職業跑道，而特地設計提供一個好事業供大家有更多的職場選項。而飲水機、淨水機的技術和知識全程事業轉移，或是全程輔導成功，以後幫助個人事業的成長，是我們著這本書的原意。因為吾人確信，產品技術和銷售知識充分具備，必能創造意想不到的大量財富；因吾人所經營的飲水機淨水設備租用事業體，長期以來也給本租用事業團隊，帶來豐碩大量的財富。所以有鑑以目前時空環境變遷的強烈的需要與需求，特地著這本《超級銷售致富祕訣》一書以供大家研讀。

祝大家心想事成、處處發大財。謝謝大家！

超級行銷講師　**徐錦坤**

井易飲水機租用加盟事業團隊編著

2020 年 5 月 30 日

Author Order 1

I am engaged in drinking fountains. The water purifier business has been around for 30 years. Now it is the head of the Jingyifeng water machine rental franchise business team. Based on the current time and space environment, there are many changes. There are many people in the society. Or many young people. Or many middle-aged people. Want to switch career track. Specially designed to provide a good career. For everyone to have more career options. And the water dispenser. The technology and knowledge of the water purifier. The whole career transfer. Or the success of the whole counseling. Help the growth of personal career in the future. This is the original intention of this book. Because I am sure that I have sufficient product technology and sales knowledge. I will be able to create an unexpectedly large amount of wealth. Because I rent a business of water purification equipment for surcharge machines. I have also given this book for a long time. Renting a business team. Brings a lot of wealth. Therefore, taking into account the strong needs and needs of the current time and space environment changes, I specially wrote this book on the secret of super sales to get rich. For everyone to study. I wish you all the best. Make a fortune everywhere. Thank you.

Super lecturer, **Xu Jinkun**

Jing Yishuang water machine hired by the franchise team.

May 30, 2020.

作者序二

現今社會個人要創業，不比經濟剛起飛時期的機會那麼多，而且需要大筆資金。有鑑於此而出本書，能幫助有志者轉動腦筋、激發鬥志與潛能，讓靠爸、媽寶、宅女族能自立自強，對社會有所貢獻，所以提供本書讓大家研讀。感謝親友團的支持，願所有接觸本書的讀者，對健康、經濟、家庭，都有所幫助，不要忘了回饋社會，多說好言，多做善事，達成禮運大同篇敘述的境界。謝謝大家！感恩大家！

超級行銷講師　徐錦坤

2020 年 7 月 8 日

Author Order 2

Today's society individuals want to start a business. Not so many opportunities than when the economy just took off. And it needs a lot of money. In view of this. And this book. Can help aspiring people. Turn your brain. Stimulate fighting spirit and potential. Let rely on dad. Ma Bao. House girl family. Can be self-reliant. Contribute to society. So provide this book. Let everyone study. Thank you for your support. I hope all readers who come into contact with this book. For health. economic. family. All help. Don't forget to give back to the society. Say more words. Do more good deeds. Reached the state of narrative in the Datong Litong. thank you all. Thanks everyone.

Super Marketing Lecturer, **Xu Jinkun Author**

July 8, 2020

講師序

潛能開發本質是腦力開發，有五層次開發：第一層次——知識更新；第二層次——技能開拓；第三層次——思維創新；第四層次——觀念轉變；第五層次——潛能開發。一個編織夢想、學習新知識、學習新技能、創新點子，甚至轉變成良好態度與生活型態，都永遠不嫌遲，太老、太年輕或者性別不適。開創好生活雖然不容易，但卻能帶給你無窮的回報。你無法使時光停止，但可以停止消極的悲觀思想。立即開始運用自己的潛能，得到你想要的，得到你想追尋的。

超級講師　**張世周**

2020 年 5 月 30 日

Preface

The essence of potential development is mental development. There are king-level development... first level... knowledge update. Second level... skill development third level... thinking innovation fourth level... concept change fifth level... potential development. Aweave Dreams. Learning new knowledge. Learning new skills. Innovating ideas. Even turning into a good attitude and lifestyle. It's never too late. Too old. Too young. Or gender discomfort. Creating a good life is not easy. But it can bring Give you endless rewards. You can't stop time. But you can stop negative pessimism. Start using your potential immediately. Get what you want. Get what you want to pursue.

Super Lecturer, **Zhang Shizhou**

May 30, 2020

目錄

A.365 祕訣 365 Secret

A1：好的人際關係，是會一帶來龐大的財富和快樂的人生，只要你相信它，它一定就會實現的這是真的，這是信念.也是成功的必備要件，切記。

A：發想…… B：目標…… C：願景……

A1: A good interpersonal relationship will bring huge wealth and a happy life. As long as you believe in it, it will be realized. This is true. This is belief. It is also a necessary requirement for success.

A: Think... B: Objective... C: Vision...

A2：一個好的客戶，一個好的貴人，可以給你帶來巨大的財富和巨大的事業發展，切記。

A：發想…… B：目標…… C：願景……

A2: A good customer. Agood nobleman. Can bring you great wealth and huge career development. Remember.

A: Think... B: Objective... C: Vision...

A3：不要怕改變，改變不是不好，凡是要去嚐試看看，這一條路走不通，換一條路走看看，總是有一條路會通，韌性，毅力，信心，決心，是走通的關鍵所在，切記。

A：發想…… B：目標…… C：願景……

A3: Don't be afraid to change. Change is not bad. Anyone who wants to try it out. This road does not work. Take another look. There is always a way. The nature. perseverance. confidence. determination. Is the key to getting through. Remember.

A: Think... B: Objective... C: Vision...

A4：機會要自己去找，機會不會找你，所以成功就在機會裡，切記。

A：發想…… B：目標…… C：願景……

A4: Opportunities must be found by yourself. Opportunity will not find you. So success lies in opportunity. Remember.

A: Think... B: Objective... C: Vision...

A5：好的人緣，就有好的人脈，好的生意，所以要建立好人緣，成功就在好人緣裡，切記。

A：發想…… B：目標…… C：願景……

A5: Good popularity. There are good connections. Good business. So we must build good people. Success lies in good relationships. Remember.

A: Think... B: Objective... C: Vision...

A6：要主動去認識經營者，大老闆，大經理人或企業負責人，或決策者，這樣成功的機率，就會大大的增加，切記。

A：發想…… B：目標…… C：願景……

A6: To take the initiative to meet the managers. big boss. Big manager or person in charge of the enterprise. Or decision maker. The probability of such success. Will greatly increase. Remember.

A: Think... B: Objective... C: Vision...

A7：掌握機會，就是掌握財富，掌握財富，也就是掌握你未來的人生，是財富成敗的關鍵所在，切記。

A：發想…… B：目標…… C：願景……

A7: Grasp the opportunity. Is to master wealth. Master wealth. That is to control your future life. Is the key to the success or failure of wealth. Remember.

A: Think... B: Objective... C: Vision...

A8：做決策，做決定，做方向，做投資，要符合實際面，現實面，絕不能情緒化，暴躁化，更不能太感性，否則會失敗，會一敗塗地，切記。

A：發想…… B：目標…… C：願景……

A8: Make a decision. Make a decision. Direction. Make an investment. To be in line with reality. Realistic. Must not be emotional. Irritability. Not too emotional. Otherwise it will fail. Will be defeated. Remember.

A: Think... B: Objective... C: Vision...

A9：改變生活，改變財物，就是改變命運致富，也就是走向未來富裕的人生，切記。

A：發想…… B：目標…… C：願景……

A9: Change life. Change property. Is to change fate and become rich. That is, toward a wealthy life in the future. Remember.

A: Think... B: Objective... C: Vision...

A10：要好好的去經營大老闆，或去經營經理人或大經理人，採購者或大採購者，或決策者，或大決策者，或企業負責人，好好建立良善的互動關係，切記。

A：發想…… B：目標…… C：願景……

A10: To run thc big boss well. Or go to the manager or big manager. Buyers or large buyers. Or decision maker. Or the big decision maker. Or the person in charge of the enterprise. Establish a good interactive relationship. Remember.

A: Think... B: Objective... C: Vision...

A11：多一個行銷動作，就多一個生意的機會，多認識一個客戶，就多一個生意的機會，所以多一個客戶，就多一個生意成交的機會，所以務必具有這樣的行銷特質和概念，方可行銷順利成功，切記。

成功是靠有準備的，成功是靠有機運的，成功是靠有責任心的，成功是靠有實力的。

Success depends on preparation. Success depends on luck. Success depends on responsibility. Success depends on strength.

A：發想…… B：目標…… C：願景……

A11: One more marketing action. Just one more business opportunity. Know one more customer. Just one more business opportunity. So one more customer. There is one more opportunity for business transactions. So be sure to have such marketing characteristics and concepts. The market can be successfully sold successfully. Centuries.

A: Think... B: Objective... C: Vision...

A12：多一點笑容與人相處，就多一點生意的機會，多一點付出，就多一點生意的機會，多一點幫忙，就多一點生意的機會，多一點關心，就多一點生意的機會，成敗的關鍵，就在這一點，這就是推銷，行銷的技巧和態度技巧，而態度和技巧是決定生意成敗的一切，切記。

A：發想…… B：目標…… C：願景……

A12: Get a little smile and get along with people. Just a little more business opportunities. Pay more. Just a little more business opportunities. Alittle more help. Just a little more business opportunities. Be more concerned. Just a little more business opportunities. The key to success or failure. At this point. This is marketing. Marketing skills and attitude skills.

A: Think... B: Objective... C: Vision...

A13：改變消費模式，改變推銷模式，改變行銷模式，改變群體接觸模式消費思維模式，推銷思維模式，行銷思維模式，切記。

A：發想…… B：目標…… C：願景……

A13: Change consumption patterns. Change the sales model. Change the marketing model. Change the group contact mode consumption thinking mode. Marketing thinking mode. Marketing thinking mode. Remember.

A: Think... B: Objective... C: Vision...

窮則變，轉彎，變則通。

The poor is the way. The turn. The way is the way.

A14：在工作上，在事業上，在人生旅途上，碰到困難或不如意，失敗或落魄或失意，事不順遂，萬萬不能灰心喪志，因為這是驅使你前進成功的最好動力，切記。

A：發想…… B：目標…… C：願景……

A14: At work. In career. On the journey of life. Encountered difficulties or unsatisfactory. Failure or disorientation or frustration. The style is not smooth. Never be discouraged. Because this is the best motivation to drive you forward to success. Remember.

A: Think... B: Objective... C: Vision...

A15：一個人要有理想，有目標，有願景，要用畢生的智慧，用才智，用智慧去完成目標和理想，和願景，這樣人生就會變的不一樣，變的有活力，有希望，有意義，要記住。

A：發想…… B：目標…… C：願景……

A15: One must have ideals. have goal. Have a vision. Use the wisdom of life. Use wisdom. Use wisdom to accomplish goals and ideals. And vision. Then life will be different. Change vitality. has hope. Significant. must remember.

A: Think... B: Objective... C: Vision...

A16：所有的業務工作，生意工作，都必須具備有，超前的佈局，佈署，和超前的規劃藍圖，這樣方可有亮眼的績效，而這是業務生意成敗的關鍵所在，要記住。

A：發想…… B：目標…… C：願景……

A16: All business work. Business work. Must have all. Advance layout. Deployment. And advanced planning blueprints. In this way, you can have bright performance. This is the key to the success of the business. must remember.

A: Think... B: Objective... C: Vision...

走過的路，走過的經驗，就是你的資產。

The road traveled. The experience traveled is your asset.

A17：在工作上，在業務上，在銷售上，在事業上，要確實落實基本工作，那麼就會產生差異化，區隔化，那麼銷售績效，自然就會浮現，切記。

A：發想 B：目標 C：願景……

A17: At work. In business. On sales. In career. We must implement the basic work. Then there will be differentiation. Segmentation. Then sales performance. Will emerge naturally. Remember.

A: Think... B: Objective... C: Vision...

A18：在生活，在事業中，有一點壓力，並不是不好，而是要把壓力轉化為生活，事業的動力，使生活事業更進步更精進，更豐富，更多彩多姿，創造富裕美好的人生，要記住

A：發想…… B：目標…… C：願景……

A18: In life. In the cause。

Alittle pressure. Not bad. It's about turning stress into life. Motivation of career. Make life and progress more advanced and more advanced. More abundant. More colorful. Create a rich and beautiful life. must remember.

A: Think... B: Objective... C: Vision...

Al9：出這本書的原意，是在提供我多年在業務上的經驗和行銷上的點點滴滴的經驗，潛能訓練的歷程和經驗，供大家參考，和大家分享，希望全心全力鼓舞你，激發你潛在的能量，正向的動力，使你更有戰鬥力，更有在人生裡面，更有競爭力，切記。

A：發想…… B：目標…… C：願景……

Al9: The original intention of this book. Is to provide my years of business experience and bit by bit experience in marketing. History and experience of potential training. For reference. Share with you. Hope to inspire you wholeheartedly. Inspire your potential energy. Positive power. Make you more combative. More in life. is more competitive.

Remember.
A: Think... B: Objective... C: Vision...

A20：在生活上，在業務上，生意上，在事業上，要確定一個不變的定位，定律，和遵循的模式，那就是今天要比昨天好和進步，而明天要比今天更好，更進步，更有實力，更幸福更充實，不必也不用跟別人比，天天要審視自己有沒有比過往更進步，這樣就可以了，只要今天比過往更進步，更好，就對了，要記住。
A：發想…… B：目標…… C：願景……
A20: In life. In business. In business. In career. To determine a constant positioning. law. And compliance mode. That is, today is better and better than yesterday. Tomorrow is better than today. More progress. More powerful. Happier and more fulfilling. No need to compare with others. Every day, we must examine whether we have made more progress than in the past. that's it. As long as today is better than ever. better. Right. must remember.
A: Think... B: Objective... C: Vision...

A21：在生活上，在工作上，在創業上，遇到困難，不要一下就說，辦不到，這個字眼，要把困難轉化為挑戰，為目標，這樣就容易多，要把困難化為容易做到的心態，心態不要一直侷限困難這兩個字，要把它當作挑戰來做，來執行。
A：發想 B：目標 C：願景……
A21: In life. at work. In entrepreneurship. Encounter difficulties. Don't just say it. Can't do it. The word. We must turn difficulties into challenges. For the goal. This is much easier. We must turn difficulties into an easy-to-do mentality. Don't limit your mind to the word difficult. Do it as a challenge. To execute.
A: Think... B: Objective... C: Vision...

A22：專注和專業，是工作事業的基石，一步一腳印，實幹實做是工作事業的基石，改變和創意是工作事業的基石，要保持隨時隨地與時俱進的心，盡心盡力發展工作和實現事業的藍圖。

A：發想…… B：目標…… C：願景……

A22: Annotation and dedication. Is the cornerstone of work. Step by step. Hard work is the cornerstone of work. Change and creativity are the cornerstones of work. To keep pace with the times, anytime, anywhere. Dedicated to the development of work and the realization of the blueprint.

A: Think... B: Objective... C: Vision...

A23：不要怕改變，改變不是不好，凡是要去嚐試看看，這一條路走不通，轉個彎換一條路走，總是有一條路會通，韌性，毅力，信心，是走得通的關鍵所在。

A：發想…… B：目標…… C：願景……

A23: Don't be afraid to change. Change is not bad. Anything to try and see. This road doesn't work. Take a turn for another road. There is always a way to pass. toughness. perseverance. confidence. The key is to make it work.

A: Think... B: Objective... C: Vision...

A24：好的人際關係，是會帶來龐大的財富和實質有成就感的人生，只要你相信它，它一定會實現的，這是真的，這是奮鬥的信念，也是富裕成功的要件。

A：發想…… B：目標…… C：願景……

A24: Good interpersonal relationships. It is a life that will bring huge wealth and substantial sense of accomplishment. As long as you believe it. It will be realized. This is real. This is the belief of struggle. It is also essential for FE's success.

A: Think... B: Objective... C: Vision...

A25：一樣的笑容，一樣的不笑容，一樣的服務態度好，一樣的銷售態度好，一樣的服務態度不好，試想購買者，消費者，會導向誰購買，答案是肯定的，笑容好，服務好，態度好，自然好業績就浮現，笑容不好，服務不好，態度不好，自然生意業績就差，這是不變的行銷定律。

A：發想…… B：目標…… C：願景……

A25: The same smile. The same does not smile. The same service attitude. The same sales attitude is good. The same service attitude is not good. Imagine the buyer. consumer. Who will lead the purchase. The answer is yes. Good smile. good service. Good attitude. Naturally good performance emerges. Smile is not good. Service is not good. Bad attitude. Natural business performance is poor. This is the constant law of marketing.

A: Think... B: Objective... C: Vision...

A26：思維，要把自己定位為思想家，實踐家，實業家，行動家，創意家，教育家，潛能訓練家，創業家，社會公益服務家。

A：發想…… B：目標…… C：願景……

A26: Thinking. We must position ourselves as thinkers. Practice at home. Industrialist. Activist. Creative home. Educator. Potential trainer. Entrepreneurs. Social welfare service home.

A: Think... B: Objective... C: Vision...

A27：推銷及銷售，銷售員笑容態度要好，你將會得到預想不到的好收成，服務員笑容態度要好，你將會得到預想不到的好業績，人際關係要好，你將會得到預想不到的成長及收入，

A：發想…… B：目標…… C：願景……

A27: Promotion and sales. The salesman smiles better. You will get an unexpectedly good harvest. The waiter smiled better. You will get unexpectedly good results. Interpersonal relationships are better. You

will get unexpected growth and income.
A: Think... B: Objective... C: Vision...

A28：愛客戶，尊重客戶，開心的服務客戶的細節，是決定客戶是否會再度來消費或購買的關鍵所在，親切，關心，用心，處處替客戶著想，隨時隨地站在客戶的立場，情景著想，充分實踐銷售員的親和力，這是決定客戶是否會再來購買和消費的關鍵所在，
A：發想…… B：目標…… C：願景……
A28: Love customers. Respect customers. Details of happy customer service. It is the key to decide whether the customer will come to consume or buy again. kind. care. Attentively. Think for customers everywhere. Stand at the customer's stand anytime, anywhere. Think about the situation. Fully implement the affinity of the salesperson. This is the key to determining whether customers will come to buy and consume again.
A: Think... B: Objective... C: Vision...

A29：在工作上，在業務上，在推銷上，在銷售上，在事業上，要確確實實落實基本工作，那麼差異化，區隔化的業績，自然就會浮現。
A：發想…… B：目標…… C：願景……
A29: At work. In business. On sales. On sales. In career. We must really implement the basic work. Then differentiate. Segmentation of performance. Will emerge naturally.
A: Think... B: Objective... C: Vision...

A30：改變對生命的態度，成功翻轉自己的人生，充實自己的生命態度，成功翻轉自己的人生，豐富自己的生命態度，成功翻轉自己的人生。

A 發想 B：目標 C：願景

A：發想…… B：目標…… C：願景……

A30: Change your attitude towards life. Successfully turned his life. Enrich your life attitude. Successfully turned his life. Enrich your life attitude. Successfully flip your own life.

A: Think... B: Objective... C: Vision...

A31：一個人往生命奮鬥的旅程中，遇到挫折或失敗，並不可怕，最怕的是遇到挫折或失敗，而喪失了向前奮鬥的熱情。

A：發想…… B：目標…… C：願景……

A31: A person's journey to life. Encountered setbacks or failures. And terrible. The most feared is encountering setbacks or failures. And lost the enthusiasm to fight forward.

A: Think... B: Objective... C: Vision...

A：32 思維，願景，要不斷的提昇行銷，銷售，企業，事業的正向能量，使行銷，銷售，企業，事業的正向能量更好，更欣欣向榮。

A：發想…… B：目標…… C：願景……

A: 32 thinking. Vision. To continuously improve marketing. Sales. enterprise. Positive energy of career. Make marketing. Sales. enterprise. The positive energy of the cause is better. More prosperous.

A: Think... B: Objective... C: Vision...

A33：工作，生意，事業，企業是用努力和時間所累積的成果，是用努力和時間所堆積的實力，而是用努力和實力所累積的後盾，有了雄厚的後盾，所有的一切，就會輝煌騰達。

A：發想…… B：目標…… C：願景……

A33: Work. business. cause. The enterprise is the result of hard work and time accumulation. It is the strength accumulated through hard

work and time. It is backed up by effort and strength. With strong backing. Everything. It will be brilliant.
A: Think... B: Objective... C: Vision...

A34：信念加堅持，是追求夢想的原動力，堅定自己的信念，堅持自己的目標，勇往直前，終究會達到自己想要的目標和願景。
A：發想…… B：目標…… C：願景……
A34: Ten beliefs. Is the driving force behind the dream. Strengthen one's faith. Stick to your goals. Go forward bravely. After all, you will achieve your desired goals and vision.
A: Think... B: Objective... C: Vision...

A35：做工作，做生意，做事業，要有一點冒險的精神，往前進的概念和決心，不然你將會一事無成，如果你今天還是原地踏步，如果你今天還是不努力，明天你就將會過苦難的日子，過苦難的生活，如果你今天好好努力，明天你就將會過繽紛多彩的生活。
A：發想…… B：目標…… C：願景……
A35: Do work. Do business. Do a career. Be a little adventurous. Concept and determination to move forward. Otherwise you will achieve nothing. If you still stand still today. If you still don't work hard today. Tomorrow you will live in misery. Live a miserable life. If you work hard today. Tomorrow you will live a colorful life.
A: Think... B: Objective... C: Vision...

A36：在工作上，在生意上，在事業上，要堅持專注和專業，用心研發和發展，直到工作，生意，事業，研發，發展至質巔峰。
A：發想…… B：目標…… C：願景……

校長，企業家，領導人，就像花圃的園丁，天天灌溉每一顆種籽，天天照顧每一顆種籽（員工），快快成長，健康壯大。
Principal. Entrepreneur. Leader. Just like the gardener of the flowerbed. Irrigating every seed every day. Taking care of every seed every day（employee）. Growing fast. Healthy and strong.

A36: At work. In business. In career. We must persist in giving attention and dedication. Careful research and development. Until work. business. cause. R & D. Development to the pinnacle of quality.
A: Think... B: Objective... C: Vision...

A37：做工作，做生意，做事業，朋友與朋友和夥伴與夥伴要攜手連結，攜手創造，啟動大家的期望和未來，創造多贏的大局面，創造多贏的大未來。
A：發想…… B：目標…… C：願景……
A37: Work. Do business. Do a career. Friends and friends and partners and partners should join hands together. Create together. Start everyone's expectations and future. Create a win-win situation. Create a win-win big future.
A: Think... B: Objective... C: Vision...

A38：做生意，還是需要大量的廣結善緣，八面玲瓏運轉，還是需要大量的推銷推廣，唯有持續不斷的大力推廣，方可生意興隆，收穫，成長，成功。
A：發想…… B：目標…… C：願景……
A38: Doing business. Still need a lot of good fortune. Exquisite operation in all directions. Still need a lot of sales promotion. Only continuous and vigorous promotion. Only business is booming. reward. growing up. success.
A: Think... B: Objective... C: Vision...

A39：努力準備實力，機會是留給有準備的人們，努力，用心，打拼，平凡人，也能創造出不平凡的精彩人生。
A：發想…… B：目標…… C：願景……
A39: Strive to prepare for strength. Opportunity is reserved for people who are prepared. Work hard. Attentively. Hard work. normal people. It can also create an extraordinary life.

只要肯上進，肯努力，貧窮也能變富有。
As long as you are willing to make progress, be willing to work hard, and poverty can become rich.

A: Think... B: Objective... C: Vision...

A40：創意，創新，執行力，行動力，冒險，膽識做規劃和決策，必有一番事業新氣象，必有一番事業新成就。
A：發想…… B：目標…… C：願景……
A40: Creativity. Innovation. Executive power. Action force. adventure. Dare to make plans and decisions. There will be new signs of career; There must be new achievements in the cause.
A: Think... B: Objective... C: Vision...

A41：人們只要打拼，立即提升生活品質，人們只要努力打拼，立刻增進生命新價值，美好的未來，立刻就會浮現。
A：發想…… B：目標…… C：願景……
A41: People just have to work hard. Improving the quality of life immediately. People just work hard. Improving the new value of life immediately. glorious future. It will appear immediately.
A: Think... B: Objective... C: Vision...

A42：一個人脈一條路，還有後面無限的路，少一個人脈，少一條路，還有後面將無路可走，路越走越窄，人生必將好無希望，路要越走越寬越遠，人生就會變的多彩多姿。
A：發想…… B：目標…… C：願景……
A42：One connection and one road. There is an infinite way behind. One less connection. One less road. There will be nowhere else to go. The road is getting narrower. Life is bound to be hopeless. The road must go wider and wider. Life will be colorful.
A: Think... B: Objective... C: Vision...

只要肯用心，肯下功夫，肯學習，處處是機會。
As long as you work hard. You work hard. You learn. Everywhere is an opportunity.

A43：當你意志堅定，志向堅定，信念堅定，將無人可以左右你的堅定，志向，信念，勇往直前，邁向勝利成功之路。

A：發想…… B：目標…… C：願景……

A43: When you are determined. Ambitious. Conviction. No one can control your firmness. ambition. belief. Go forward bravely. On the road to victory and success.

A: Think... B: Objective... C: Vision...

A44：人生機會無數多，每天都是機會，處處都是機會，每個機會都要緊緊的把握住，把握住就不一定有機會，若不握住，根本就沒機會，這是老天給每一個人同等的機會，就看你會不會把握機會，不把握住機會，肯定是失敗者，把握住機會是成功者，無需等待，也就是說實現夢想。

A：發想…… B：目標…… C：願景……

A44: There are countless opportunities in life. Every day is an opportunity. Opportunities are everywhere. Hold every opportunity tightly. There is not necessarily a chance to seize it. If not hold. There is no chance at all. This is God giving everyone the same opportunity. It depends if you take the chance. Do not seize the opportunity. Definitely a loser. Seize the opportunity is the winner. No need to wait. That is to realize the dream.

A: Think... B: Objective... C: Vision...

A45：只要肯努力，無論身在何處，遲早都會發現你的好，你的優質，你的奮鬥價值。

A：發想…… B：目標…… C：願景……

A45: Just be willing to work hard. No matter where you are. Sooner or later I will find you good. Your quality. The value of your struggle.

A: Think... B: Objective... C: Vision...

A46：勤奮的人，要彎下腰，低下頭，好好推銷推廣，深信一定有好成績，好績效，持續一直做，總有一天，一定會有結果。

A：發想…… B：目標…… C：願景……

A46: Hardworking person. To bend down. Head down. Good sales promotion. Convinced that there must be good results. Good performance. Holding performance has been done. Someday. There will be results.

A: Think... B: Objective... C: Vision...

A47：時間可以成就很多的事情事物，浪費時間，也可以失去很多的事物事情，所以把握時間是創造人生成功的關鍵所在。

A：發想…… B：目標…… C：願景……

A47: Time can accomplish many things. waste time. You can also lose a lot of things. So grasping the time is the key to the success of life.

A: Think... B: Objective... C: Vision...

A48：不要小看自己的能力，也不要時時放大自己的能力.自己的優點，也不要時時放大自己的存在時時要謙虛待人處事，只要細心學習，和耐心奮鬥前進，必有好收穫。

A：發想…… B：目標…… C：願景……

A48: Don't underestimate your ability. Don't amplify your ability from time to time. Your own advantages. Don't enlarge your deposit from time to time. Always be modest in everything. Just study carefully. Strive forward with patience. There will be good harvests.

A: Think... B: Objective... C: Vision...

A49：一個公司經營者，員工或執行者，都要把客戶再小的事，當作公司的大事，來回應，來服務，來完成，

A：發想…… B：目標…… C：願景……

A49: A company operator. Employee or executor. Have to put customers in small things. As a major event for the company. To respond. To serve. To be done.
A: Think... B: Objective... C: Vision...

A50：當你人生面對重大的打擊挫折，重大的困難時，首先要放下心來，勇敢去面對，去解決，去接受，去改變，方可一切圓滿，一切順利。
A：發想…… B：目標…… C：願景……
A50: When you face a big fight in your life. frustration. In case of major difficulties. First of all, rest assured. Face it bravely. To solve. Go accept. To change. All is complete. all the best.
A: Think... B: Objective... C: Vision...

A51：能力和實力，隨時隨地要強化自己能力和實力，成功有一些是要靠機緣，但最重要的還是要靠努力和實力，所以隨時隨地要強化自己的生活能力和事業實力。
A：發想…… B：目標…… C：願景……
A51: Ability and strength. Strengthen your ability and strength anytime, anywhere. Some success depends on chance. But the most important thing is to rely on hard work and strength. Therefore, we must strengthen our living ability and career strength anytime, anywhere.
A: Think... B: Objective... C: Vision...

A52：一個人，對自己不看不足，只看自己變得如何，敢衝，敢打拼，敢改變，敢創新，創造不同的人生境界。
A：發想…… B：目標…… C：願景……
A52: One person. Don't look down on yourself. Just look at how you have changed. Dare to rush. Dare to work hard. Dare to change. Dare to innovate. Create different realms of life.

A: Think... B: Objective... C: Vision...

A53：簡單的事，只要重複的做，做到完美極致，永不放棄，你就是一個成功的人。
A：發想…… B：目標…… C：願景……
A53: Simple things. Just do it repeatedly. Be perfect. never give up. You are a successful person.
A: Think... B: Objective... C: Vision...

A54：人只要肯努力，有信心，有信念，任何事，萬事都有可能有收穫，會成功的。
A：發想…… B：目標…… C：願景……
A54: As long as people are willing to work hard. Have confidence. Have faith. anything. Everything is possible to gain. Will succeed.
A: Think... B: Objective... C: Vision...

A55：被動創造商機，被動創造未來，主動創造商機，主動創造未來，被動是等待，主動是創造先機，掌握機會，是做事成功的要素，也是成功的關鍵所在。
A：發想…… B：目標…… C：願景……
A55: Passive creation of business opportunities. Passively create the future. Actively create business opportunities. Take the initiative to create the future. Passive is waiting. Initiative is the first opportunity to create. Seize the opportunity. Is the key to success. It is also the key to success.
A: Think... B: Objective... C: Vision...

A56：觀念和心態，做人做事，隨時要調整你的心態和觀念，隨時要調整你的心態，方能改變你人生機遇和命運。

A：發想…… B：目標…… C：願景……

A56: Ideas and mentality. Doing things. Always adjust your mindset and ideas. Always adjust your mindset. Only change your life opportunity and destiny.

A: Think... B: Objective... C: Vision...

A57：如何經營單一人脈，團體人脈，首先要 A.合群 B.熱誠 C.熱情 D.熱心 E.服務 F 做感動人的事物，然後做事，這樣就能圓融圓滿做好人際關係。

A：發想…… B：目標…… C：願景……

A57: How to run a single network. Group connections. First of all, A: He Qun B, enthusiasm C, enthusiasm D, enthusiasm E, service F to do moving things. After doing things. In this way, interpersonal relationships can be well established.

A: Think... B: Objective... C: Vision...

A58：任何一件事物，或所有事物的達成，完成，必須所有一群人共同支援，共同輔助，共同信念，共同目標，才能成就所有的事和物，這就是團隊合作的概念。

A：發想…… B：目標…… C：願景……

A58: Anything. Or the achievement of everything. carry out. Must be supported by all groups. Joint assistance. Common belief. Same target. In order to accomplish all things and things. This is the concept of teamwork.

A: Think... B: Objective... C: Vision...

A59：世界上幾乎成為專業的人士，成為專業專家，或成為達人，之所以能成功，幾乎都靠堅忍，執著，不放棄這幾個字，而成為專家，成為專業達人。

A：發想…… B：目標…… C：願景……

A59: The people in the world have almost become unemployed. Become a professional. Or become a master. The reason for success. Almost depends on perseverance. Perseverance. Don't give up these words. Become an expert. Become a professional.
A: Think... B: Objective... C: Vision...

A60：做業務員，做業務工作，遊說客戶，始終沒有放棄的權利，也沒有灰心的權利，只要往前衝的責任，全心全力達成目標的使命感。
A：發想…… B：目標…… C：願景……
A60: Be a salesman. Do business work. Lobbying customers. Never give up the right. There is no right to be discouraged. Just move forward with the responsibility. Asense of mission dedicated to achieving the goal.
A: Think... B: Objective... C: Vision...

A6l：失敗是成功之母，但失敗比成功更重要，因從失敗中學到往前沒有的歷練和經驗，從歷練和經驗中，走向成功的概念。
A：發想…… B：目標…… C：願景……
A6l: Failure is the mother of success. But failure is more important than success. Because from the failure to learn less experience and experience. From experience and experience. Success concept.
A: Think... B: Objective... C: Vision...

A62：多一份笑容，就多一份業務量，多一筆生意量，多一份推銷，就多一份業務量，多一筆生意量，多一份推銷，就多一份生意量。
A：發想…… B：目標…… C：願景……
A62: One more smile. Just one more business volume. More business. One more sale. Just one more business volume. More business. One more sale. Just one more volume of business.

走出舒適圈，突破舒適圈，改變舒適圈。
Out of the comfort zone. Break through the comfort zone. Change the comfort zone.

A: Think... B: Objective... C: Vision...

A63：人們時時要自我加油，自我訓練，培養出時時打拼，時時奮鬥的習慣，培養出打不倒的決心，打不倒的勇氣，打不倒意志

A：發想…… B：目標…… C：願景……

A63: People always have to cheer on themselves. Self-training. Work hard from time to time. The habit of struggling all the time. Cultivate the determination not to beat. Unstoppable courage. Can't beat the will

A: Think... B: Objective... C: Vision...

A64：生意靠一點點的支援公關費用，有助於業務拓展，也有助於生意發展面，唯有這樣，或許這樣，生意或許會自然而然的欣欣向榮。

A：發想…… B：目標…… C：願景……

A64: The business depends on a little support PR costs. Help business development. It also helps in business development. Only then. Maybe so. Business may naturally flourish.

A: Think... B: Objective... C: Vision...

A：65：看別人看不到的商機，想別人想不到的商機，做別人不要做的商機，這個就是商機差異化，商機區隔化，這個就是你的優質，跟別人不一樣的特點和優秀。

A：發想…… B：目標…… C：願景……

A: 65: See business opportunities that others cannot see. Think of business opportunities unimagined by others. Do not do business opportunities for others. This is the differentiation of business opportunities. Business opportunity segmentation. This is your quality. Features and excellence that are different from others.

A: Think... B: Objective... C: Vision...

A66：創業，做生意，做事業，要勇敢的去適應和接受「不一樣」這三個字，這三個字的概念和觀念很重要，這決定你往後創業，做生意，做事業創新成敗的關鍵所在。
A：發想…… B：目標…… C：願景……
A66：Entrepreneurship. Do business. Do a career. Be brave to adapt and accept the three words "not the same". The concepts and concepts of these three words are very important. This determines your future business. Do business. The key to success or failure in business innovation.
A: Think... B: Objective... C: Vision...

A67：悲觀就沒有站起來的力量，肯定自我，深信自我，勇敢接受任何的挑戰，邁力實現自我希望的第一步，邁力實踐自我理想的第一步。
A：發想…… B：目標…… C：願景……
A67: Pessimism has no power to stand up. Affirm yourself. Believe in yourself. Be brave to accept any challenge. The first step towards realizing self hope. The first step towards realizing your ideals.
A: Think... B: Objective... C: Vision...

A68：只要你花一點點時間去做思維，做發想，也許能改變你的一生，給自己一個機會，改變人生的契機，成功的契機，成敗的契機，就在這裡面。
A：發想…… B：目標…… C：願景……
A68: As long as you spend a little time thinking. Think about it. Maybe it can change your life. Give yourself a chance. Opportunity to change your life. Opportunity for success. Opportunity for success or failure. Right here.
A: Think... B: Objective... C: Vision...

A69：A 求貴 B 求財 C 求福，求貴是希望能遇到貴人的提拔和幫助，求財是期許能擁有更多金錢財物，求福是期許能得到更多的人生健康和幸福。
A：發想…… B：目標…… C：願景……
A69: A seeks wealth B seeks wealth C seeks wealth. Begging is to hope to meet the promotion and help of nobles. Begging for money is to expect to have more money. Seeking wealth is the hope to get more health and wealth in life.
A: Think... B: Objective... C: Vision...

A70：處理任何的事情，或處理任何的人，事，物，必須用漸進式的模式去完成，去達成，換言之，就是執行力要循序漸進的去執行，因時間是決解問題的關鍵所在，也是達成目標的關鍵所在。
A：發想…… B：目標…… C：願景……
A70: Handle anything. Or deal with anyone. thing. Things. It must be done in a progressive mode. To achieve. In other words. It means that the execution must be carried out step by step. Because time is the key to solving problems. It is also the key to achieving the goal.
A: Think... B: Objective... C: Vision...

A71：選擇工作，選擇事業，要選擇你所喜歡的，選擇你所熱愛的，專注投入，那麼自然而然，就能累積你的財富。
A：發想…… B：目標…… C：願景……
A71: Choose a job. Choose a career. Choose what you like. Choose what you love. Note input. So naturally. Can accumulate your wealth.
A: Think... B: Objective... C: Vision...

A72：客戶對你的喜歡，信任，肯定是你業務成功的基石，悲觀就沒有再站起來的力量和機會，悲觀也會使你失去再一次成功的契機。

A：發想…… B：目標…… C：願景……

A72: Customers like you. trust. It must be the cornerstone of your business success. Pessimism has no strength and opportunity to stand up again. Pessimism will also make you lose the opportunity to succeed again.

A: Think... B: Objective... C: Vision...

A73：用微笑認識生活，用時間累積實力，用實力堆砌事業的後質，用努力和創意，打造穩固的事業版圖。

A：發想…… B：目標…… C：願景……

A73: Know life with a smile. Accumulate strength with time. Use the strength to build the back quality of the cause. With effort and creativity. Create a solid career map.

A: Think... B: Objective... C: Vision...

A74：成功是一棵種籽，要種在打拼，勤奮，前進的旅途上，有耐心，有毅力，有決心，往前衝，永不放棄，成功就會很快的浮現在你的眼前。

A：發想…… B：目標…… C：願景……

A74: Success is a seed. To be hard at work. diligent. On the journey ahead. Be patient. Have perseverance. Determined. Rush forward. never give up. Success will soon appear in front of your eyes.

A: Think... B: Objective... C: Vision...

A75：每個客戶都很重要，你怎樣去認識一個好客戶，如何去經營一個好客戶，認識的好，經營的好，你就必將有所收穫.你就必將前無限，你失去一個好客戶，你將失去一片天，你將必失去大片財富，務必謹記在心。

A：發想…… B：目標…… C：願景……

要訓練好笑容，要訓練好相處，要訓練好合群。

To train a good smile. To train to get along well. To train a group.

A75: Every customer is very important. How do you know a good customer. How to run a good customer. Good understanding. Good management. You will have something to gain. You will have unlimited front. You lose a good customer. You will lose a day. You will lose a lot of wealth. Always keep in mind.
A: Think... B: Objective... C: Vision...

A76：一句惡言，失去整千萬，一句好言，賺進整千萬，一句好言，得一生好伙伴，好朋友，一句惡言，失去一生好伙伴，一生好朋友。
A：發想…… B：目標…… C：願景……
A76: A sad word. Lost ten million. Agood word. Earn ten million. Agood word. Have a good partner for life. good friend. Asad word. Lost a good partner for life. Agood friend for life.
A: Think... B: Objective... C: Vision...

A77：客戶是你的衣食父母，客戶是你生活的保障，客戶是你生活的貴人，有了保障，有了貴人，你的生活就會很有希望，願望自然就會順理成章的達成。
A：發想…… B：目標…… C：願景……
A77: The customer is your food and clothing parent. Customers are the guarantee of your life. Customers are the nobles of your life. With security. With nobles. Your life will be very promising. The wish will naturally come true.
A: Think... B: Objective... C: Vision...

A：78 夢想……就是擘劃出一條可長可遠的好的創業路，理想……就是把好的創業路，就必須有好的團隊，來共同合作，携手打拼，共同奮鬥，達成好的創業路，幻想……就是空想，就是沒有行動力的想法，沒有付諸行動的做，原地踏步。

人脈是致富的跳板，人脈不一定是錢脈，而是要有正確的人脈，才正是錢脈。
The network is a springboard for getting rich. The network is not necessarily the money. It is the right network. It is the money.

A：發想…… B：目標…… C：願景……
A: 78 Dream... It is to create a good long-term entrepreneurial road. Ideal... is to take a good entrepreneurial path. There must be a good team. Come to work together. Work hard together. Struggle together. Agood way to start a business. Fantasy... is fantasy. It is an idea without action. Did not put into action. No progress.
A: Think... B: Objective... C: Vision...

A79：技術加商業知識，技術就是製造出產品，商業知識，就是推銷佈局，把產品銷售出去，兩者相輔相成、相得益彰，企業才有可能大發展，所以技術和商業知識是企業是否發展的關鍵所在。
A：發想…… B：目標…… C：願景……
A79: Technology ten business knowledge. Technology is to make products. Business knowledge. It is the marketing layout. Sell the product. The two complement each other and complement each other. It is only possible for enterprises to develop. Therefore, technology and business knowledge are the key to the development of an enterprise.
A: Think... B: Objective... C: Vision...

A80：樹立銷售員的服務形象，提昇銷售員的服務要求，銷售就是王道，服務就是王道。
A：發想…… B：目標…… C：願景……
A80: Establish the service image of salesperson. Improve the service requirements of sales staff. Sales is king. Service is king.
A: Think... B: Objective... C: Vision...

A81：想像力很重要，知識也很重要，試著去找到一條致勝的道路，多讀書，多閱啟發你的智慧，知識就是力量。
A：發想…… B：目標…… C：願景……

人脈是致勝的關鍵，也是事業成功的保證。
Networking is the key to winning. It is also the guarantee of career success.

A81: Imagination is very important. Knowledge is also very important. Try to find a way to win. read more books. Read more to inspire your wisdom. knowledge is power.
A: Think... B: Objective... C: Vision...

A82：一步一腳印，千萬不要放棄任何可能成功的小細節，也千萬不要放棄可能成功的小機會，掌握細節，掌握機會，是成功的要件。
A：發想…… B：目標…… C：願景……
A82: One step at a time. Never give up any small details that may succeed. Don't give up the small chance of success. Master the details. Take advantage of opportunities. It is the key to success.
A: Think... B: Objective... C: Vision...

A83：人在做生意中，人在創業中，將會遇到千佰次的困難，挫折，沮傷，失意，灰心，喪志，隨時隨地要反思自己過往的錯誤和錯敗，而這是你成功的基石和養分，隨時檢視自己的誤判和錯誤，那些地方做的不好，不周詳，不仔細，而再出發，而這是成功的基礎和養分。
A：發想…… B：目標…… C：願景……
A83: People are doing business. People in entrepreneurship. Will encounter difficulties hundreds of times. frustration. Sad. Frustrated. discouraged. Bereavement. Rethink your past mistakes and failures anytime, anywhere. And this is the cornerstone and nutrients of your success. Check your misjudgments and errors at any time. Those places are not doing well. Not detailed. careless. And start again. And this is the basis and nutrients of success.
A: Think... B: Objective... C: Vision...

A84：創業要靠膽識，成功要靠實力，這自然界不變的定律。
A：發想…… B：目標…… C：願景……

A84: Entrepreneurship depends on courage. Success depends on strength. This constant law of nature.
A: Think... B: Objective... C: Vision...

A85：面對消費者，好的開心服務細節，不舒服的服務細節，是決定顧客是否會再來消費的關鍵所在，務必謹記。
A：發想…… B：目標…… C：願景……
A85: Facing consumers. Good happy service details. Uncomfortable service details. Is the key to determine whether customers will come back to consume. Remember.
A: Think... B: Objective... C: Vision...

A86：人必須大力的改變你的生活環境，含蓋人，事，時，地，物，改變所有不成功的因索，那麼一切自然而然就會順著改，那麼你離成功，就會愈來愈近。
A：發想…… B：目標…… C：願景……
A86: People must vigorously change your living environment. Covered people. thing. Time. To. Things. Change all unsuccessful causes. Then everything will change naturally. Then you are successful. Will get closer and closer.
A: Think... B: Objective... C: Vision...

A87：成功銷售……就是引導客戶不想購買而購買，發覺客戶本身沒有發覺的需求而購買.這是成功銷售的關鍵所在。
A：發想…… B：目標…… C：願景……
A87: Successful sales... is to guide customers to buy without wanting to buy. It is found that customers themselves do not have the need to make purchases. This is the key to successful sales.
A: Think... B: Objective... C: Vision...

A88：在工作上，在生意上經過千佰次的變革改革，經過千佰次的重挫和失敗，經過一次又一次的重挫和打擊，而這是你奠定成功的實力和基石。

A：發想…… B：目標…… C：願景……

A88: At work. After thousands and thousands of revolution reforms in business. After thousands of times of setbacks and failures. After repeated setbacks and knockouts. And this is your strength and cornerstone for success.

A: Think... B: Objective... C: Vision...

A89：做生意，做事業，無論遇到任何的困苦或困難，不要當下就認為很困苦很困難，退一步，慢一點，轉個彎，事情會不一樣的，任何的困難，任何的問題，都要想辦法去面對它，處理它，解決它，使其萬事圓融圓滿。

A：發想…… B：目標…… C：願景……

A89: Do business. Do a career. No matter what difficulties or difficulties. Don't think it's hard and difficult now. Take a step back. slower. Turn a corner. Things will be different. Any difficulties. Any problems. We must find ways to face it. Deal with it. fix it. Make everything perfect.

A: Think... B: Objective... C: Vision...

A90：事業要成功，就要敢衝，敢變，敢挑戰，不怕挑戰，這是滋潤你成功的養分，韌性，耐性是你成功的基石，敢衝敢改變敢挑戰，是事業成功惟一的機會。

A：發想…… B：目標…… C：願景……

A90: Career must succeed. Dare to rush. Dare to die. Dare to challenge. Not afraid of challengcs. This is the nutrient that nourishes your success. toughness. Patience is the cornerstone of your success. Dare to dare to change and dare to challenge. It is the only opportunity for career success.

A: Think... B: Objective... C: Vision...

A91：在事業上，從挫折中累積翻轉的能量，從重擊中，失敗中堆疊再出發的能量和勇氣，等待機會，相信這樣，絕對能翻轉成功的人生。
A：發想…… B：目標…… C：願景……
A91: In business. Accumulate flip energy from frustration. From the heavy fight. The energy and courage to start again in failure. Waiting for an opportunity. I believe so. It can definitely turn over a successful life.
A: Think... B: Objective... C: Vision...

A92：想像夢想和希望的圖樣，心底有一個夢想，認真的態度是無價，認真加努力加實踐=成功。
A：發想…… B：目標…… C：願景……
A92: Picture of dreams and hopes. There is a dream in my heart. Aserious attitude is priceless. Serious ten efforts ten practice = success.
A: Think... B: Objective... C: Vision...

A93：經歷過，就變成實力，不是看到希望才堅持，而是堅持才能看到希望和未來，做事不要怕麻煩，麻煩才能學到經驗，麻煩+經驗是堆疊出你未來發展的實力。
A：發想…… B：目標…… C：願景……
A93: Experienced. It becomes strength. Not insisting on seeing hope. But perseverance can see hope and future. Don't be afraid of trouble when doing things. It takes trouble to learn experience. Trouble + experience is to stack your strength for future development.
A: Think... B: Objective... C: Vision...

A94：因為有想，才有改變，因為有改變，才會有作為，敢做，敢衝，想法多，點子多，是創意成功要素。

A：發想…… B：目標…… C：願景……

A94: Because I have thoughts. Only changed. Because there is a change. Will do something. Dare to do it. Dare to rush. Many ideas. Many ideas. It is an element of creative success.

A: Think... B: Objective... C: Vision...

A95：成功要件……就是你要尋找機會，而不是機會尋找你，是創造機會，而不是等待機會。

A：發想…… B：目標…… C：願景……

A95: The key to success... is that you are looking for opportunities. Not an opportunity to find you. Is to create opportunities. Instead of waiting for opportunities.

A: Think... B: Objective... C: Vision...

A96：堅持理想，堅持目標，勇敢前進再前進，創造不可能的可能，愈挫愈勇，乘風破浪，只要肯用心，肯吃苦，肯努力，成功的機會自然而然就會浮現。

A：發想…… B：目標…… C：願景……

A96: Stick to idcals. Stick to the goal. Go forward bravely. Create impossible possibilities. More frustrated and more courageous. Ride the wind and waves. Just be willing to work hard. endeavor. Willing to work hard. Opportunities for success will naturally emerge.

A: Think... B: Objective... C: Vision...

A97：做生意，做人，做事，做任何一件事物，都務必前面一句禮貌，後面一句「謝謝」為夢想而構思，為成長而打拼，這樣就了你人生事業版圖。

A：發想…… B：目標…… C：願景……

有關係=沒關係，沒關係，就是要去找關係，找不到關係，就是要去拉關係，拉不到關係，就是有關係。

It does not matter. It does not matter. It does not matter. It is to find a relationship. It does not find a relationship. It is to pull a relationship.

A97: Doing business. Be a man. work. Do anything. Always be polite in the previous sentence. The latter sentence "Thank you" was conceived for the dream. Work hard for growth. This will give you a career map.
A: Think... B: Objective... C: Vision...

A98：不要怕挫敗，勇於走出舒適圈，創新跳出舊有的思維框架，新思維，新創新，大步向前推進，這是成功的要件。
A：發想…… B：目標…… C：願景……
A98: Don't be afraid of frustration. Courage to get out of the comfort zone. Innovation jumps out of the old thinking framework. New thought. New innovations. Stride forward. This is the key to success.
A: Think... B: Objective... C: Vision...

A99：客戶是你的衣食父母，要好好的以禮相待，客戶和朋友是你最好，最可靠的事業後盾。
A：發想…… B：目標…… C：願景……
A99: The customer is your food and clothing parent. Treat each other well. Customers and friends are your best.The most reliable career backing.
A: Think... B: Objective... C: Vision...

A100：好的客戶，好的員工，是企業永續發展最好的資產，好的朋友，好的貴人，好的人脈，也是企業發展最好的後盾。
A：發想…… B：目標…… C：願景……
A100: Good customer. Good staff. It is the best asset for the sustainable development of enterprises. Good friend. Good nobleman. Good connections. It is also the best backing for enterprise development.
A: Think... B: Objective... C: Vision...

華人獨特的人際關係：
A.認親戚 B.拉關係 C.攀交情 D.做人情 E.鑽營 F.送禮。
The unique interpersonal relationship of the Chinese:
A. Recognize relatives B. Pull relationships C. Cultivate friendship D. Do human relations E. Drill camp F. Give gifts.

A101：好的客戶，好的員工，是企業永續發展最好的資產，好的朋友，好的人脈，好的貴人.是企業發展最好的後盾，只要踏實，勤奮，肯用心，肯努力，肯創新，生命總會找到出路，勇气加堅持，不斷尋求突破事業瓶頸，營造出一個繽紛多采多姿的生活。

A：發想…… B：目標…… C：願景……

A101: Good customer. Good staff. It is the best asset for the sustainable development of enterprises. Good friend. Good connections. Good nobles. It is the best backing for the development of enterprises. Just be practical. diligent. Willing to work hard. Willing to work hard. Willing to innovate. Life will always find a way out. Courage + perseverance. Constantly seek to break through the bottleneck of the cause. Create a colorful life.

A: Think... B: Objective... C: Vision...

A102：把簡單的事情，做到最仔細最好，這就是成功極致的第一步，鎖定目標客群，經營目標客群，拓展目標客群，群策群力，這也是成功極致的第一步。

A：發想…… B：目標…… C：願景……

A102: Put simple things. Be the most careful and best. This is the first step to the ultimate success. Target the target customer group. Business target customer group. Expand the target customer base. Work together. This is also the first step in the ultimate success.

A: Think... B: Objective... C: Vision...

A103：在工作上，在生意上，在事業上，差異化……是成功極致的第一步，區隔化……也是成功極致的第一步，也是企業重要特質，更是企業重要的資產。

A：發想…… B：目標…… C：願景……

A103: At work. In business. In career. Differentiation... is the first step to the ultimate success. DiVision... is also the first step in the ultimate

先認識人，再交流，再交心，這是交朋友最佳上策。

Meet people first. Then communicate. Then make friends. This is the best way to make friends.

success. It is also an important characteristic of an enterprise. It is also an important asset of an enterprise.
A: Think... B: Objective... C: Vision...

A104：在生意上，在事業上，機會是需要自己去找的，去鋪陳的，去佈局的，生意是需要自己去開發的，去創造的，去達標的。
A：發想…… B：目標…… C：願景……
A104: In business. In career. Opportunities need to be found by yourself. Go to shop. To layout. Business needs to be developed by yourself. To create. Go to the standard.
A: Think... B: Objective... C: Vision...

A105：一個客戶，就是一個機會，就是一個希望，珍惜容戶，愛惜客戶，開發客戶群，開發市場群。
A：發想…… B：目標…… C：願景……
A105: A customer. Is an opportunity. Is a hope. Cherish Rong Ge. Cherish customers. Develop customer base. Developing market clusters.
A: Think... B: Objective... C: Vision...

A106：生意成功的關鍵在於市場的區隔化，市場的差異化，堅持自己的價值和品質，終究會穫得顧客的信賴，所謂……花兒盛開，蝴蝶自然來的自然法則。
A：發想…… B：目標…… C：願景……
A106: The key to business success is market segmentation. Differentiation of the market. Adhere to their own value and quality. After all, it will gain the trust of customers. The so-called... flowers bloom. The natural laws of butterflies come naturally.
A: Think... B: Objective... C: Vision...

A107：對工作，對事業，要有強烈的責任感和熱愛，專業感和熱情，我們不一定要很有錢，但一定要做一個值得讓人尊重，和做一個值錢的人。

A：發想…… B：目標…… C：願景……

A107: Right to work. For the cause. Have a strong sense of responsibility and love. Professionalism and enthusiasm. We don't have to be rich. But there must be something worth respecting. And be a valuable person.

A: Think... B: Objective... C: Vision...

A：108：安逸無法讓人有創意，有創新，成長必須經過無數次的挫敗和磨難及反思，方能有戰勝成功的浮現，要具備有勇於嘗試的特質，不斷的突破困難和瓶頸，不輕言放棄。

A：發想…… B：目標…… C：願景……

A: 108: Anyi cannot make people creative. There is innovation. Growth must go through countless times of frustration and suffering and reflection. Only then can the victory emerge. To have the qualities to have the courage to try. Continuous breakthroughs in difficulties and bottlenecks. Don't give up in a flash.

A: Think... B: Objective... C: Vision...

A109：一個客戶，就是一個通路，建立你第一千個的通路，這樣不但能生意通暢，也能致勝，更能致富。

A：發想…… B：目標…… C：願景……

A109: A customer. Is a pathway. Establish your thousandth channel. This will not only smooth the business. Can also win. Get richer.

A: Think... B: Objective... C: Vision...

Al10：堅持高品質，高服務概念和心態，堅持價位價值，這樣才能穩定成長，穩定發展，有一顆熱情服務的心，去服務每一個客戶，服務做到最細緻，最頂級，這就是成功極致的第一步。
A：發想…… B：目標…… C：願景……
Al10: Adhere to high quality. High service concept and mentality. Stick to price value. Only in this way can we grow steadily. steady development. Have a warm heart of service. To serve every customer. The service is the most meticulous. Top level. This is the first step to the ultimate success.
A: Think... B: Objective... C: Vision...

A111：銷售能否成功……有五個成功術，A 誠心 B 細心 C 貼心 D 耐心 E 恆心，把服務做到超出客戶的想像，極致，讓客戶感到貼心感動，這是決定銷售是否成功的關鍵所在。
A：發想…… B：目標…… C：願景……
A111: Can sales succeed... There are five successful techniques。
Asincere B careful C intimate D patience E perseverance. Make the service beyond customer's imagination. Extreme. Let customers feel intimate and moved. This is the key to the success of sales.
A: Think... B: Objective... C: Vision...

A112：只要有一顆願意服務的心，這樣生意到處都是機會，到處都是機會，到處都是客戶，這是自然不變的常態。
A：發想…… B：目標…… C：願景……
A112: As long as there is a heart willing to serve. Such business is full of opportunities everywhere. There are opportunities everywhere. Customers are everywhere. This is the normal state of nature.
A: Think... B: Objective... C: Vision...

A113：生意推銷概念.銷售概念，要以合理適當的價位來協商溝通成交，這樣生意才能做得又深又寬，長長久久。

A：發想…… B：目標…… C：願景……

A113: Business sales concept. Sales concept. Negotiate and negotiate deals at reasonable and appropriate prices. Only in this way can the business be deep and wide. Long and long.

A: Think... B: Objective... C: Vision...

A114：銷售……發廣告單，就是在種一顆種子，一顆樹苗，慢慢在發芽，成長，成長變大樹，大收穫。

A：發想…… B：目標…… C：願景……

A114: Sales... send advertisements. Is planting a seed。

Asapling. Sprouting slowly. growing up. Growing into a big tree. Great harvest.

A: Think... B: Objective... C: Vision...

A115：做生意，要求近望遠，不要一開始就捨近求遠，這樣才能順勢而成功。

A：發想…… B：目標…… C：願景……

A115: Do business. Requirements close to the horizon. Don't stay away from the start. Only in this way can we succeed.

A: Think... B: Objective... C: Vision...

A116：做生意，用熱情，微笑，親切招覽客戶，企業商機潛力將無窮大。

A：發想…… B：目標…… C：願景……

A116: Do business. With enthusiasm. smile. Kindly invitc customers. The potential of business opportunities will be infinite.

A: Think... B: Objective... C: Vision...

公司運作，決策，效益比效率更重要。

Company operation. Decision-making. Benefit is more important than efficiency.

A117：笑容，微笑，是業務與客戶在談生意時，最重要緩衝劑，也是人與人之間相處，最好的滋潤劑，這是業務人員談生意是否成功的關鍵所在，也是人與人之間是否相處融洽的關鍵所在。

A：發想…… B：目標…… C：願景……

A117: Smile. smile. When the business is talking to customers. The most important buffer. Also get along with each other. The best moisturizer. This is the key to business personnel talking about the success of the business. It is also the key to whether people get along well.

A: Think... B: Objective... C: Vision...

A118：企業做事情很快，做決策很慢，意思是說……凡是做決策前，要有縝密周詳的計劃，一旦做好決策的決定，就要去徹底的執行，徹底的實現計劃和目標。

A：發想…… B：目標…… C：願景……

A118: Enterprises do things quickly. Making decisions is slow. The idea is... every time before making a decision. Have a meticulous and detailed plan. Once the decision is made. It is going to be thoroughly implemented. Completely achieve the plan and goals.

A: Think... B: Objective... C: Vision...

A119：做生意，要給客戶最快速的服務，最快速的滿意，和最快速達成客戶的需要和交待事項。

A：發想…… B：目標…… C：願景……

A119: To do business, we must provide customers with the fastest service, the quickest intentions, and the fastest to meet the needs and accountable matters of customers.

A: Think... B: Objective... C: Vision...

好心情，好態度，好人緣，就會有好的人際關係。

Good mood. Good attitude. Good relationship. There will be good relationships.

A120：開公司，做生意，做事業，商品知識要精進熟練，商品技術要精進熟練，推銷話術要精進熟練，行銷技巧要精進熟練，方可營收和發展突飛猛進。

A：發想…… B：目標…… C：願景……

A120: Start a company. Do business. Do a career. Commodity knowledge should be refined and proficient. Commodity technology should be advanced. Promote marketing skills to be proficient. Marketing skills should be refined and proficient. Only then can revenue and development advance by leaps and bounds.

A: Think... B: Objective... C: Vision...

A121：外面市場不認識的人群，個個都是客戶，個個都不能得罪，不能放棄，得罪和放棄，就會去一大片生意，一大片榮景，地球是圓的，屋遲早都會認識，會碰到，千萬要時時牢記。

A：發想…… B：目標…… C：願景……

A121: People who do not know the outside market. All are customers. No one can offend. Can not give up. Offend and give up. Will go to a large business. Alarge area of glory. The Earth is round. Sooner or later, the house will know. Will encounter. Always keep in mind.

A: Think... B: Objective... C: Vision...

A122：時間證明，我們的努力，戰勝了貧窮，達成脫貧的目標，身體健康，生命安全，才是人生最大的財富和資產。

A：發想…… B：目標…… C：願景……

A122: Proof of time. Our efforts. Overcame greed and poverty. Achieve the goal of poverty alleviation. Healthy body. life safety. It is the greatest wealth and asset in life.

A: Think... B: Objective... C: Vision...

A123：選擇自己要走的路和理想，勇敢的走，奮發的走，努力的走，用力的走，前進再前進，不斷的前進，走向目標和理想。

A：發想…… B：目標…… C：願景……

A123: Choose your own way and ideal. Go bravely. Go hard. Go hard. Walk hard. Go ahead and go forward. Keep moving forward. Towards goals and ideals.

A: Think... B: Objective... C: Vision...

A124：生意有推廣，就會有希望.有宣傳就會有收穫.有推銷就希望，有努力就有前程。

A：發想…… B：目標…… C：願景……

A124: Business promotion. There will be hope. If there is publicity, there will be gains. If there is sales, there is hope. There is a future with hard work.

A: Think... B: Objective... C: Vision...

A125：公司要經營好，必須具備下面几個條件，A 優質的產品 B 研發產品團隊 C 行銷團隊 D 快速的維修團隊 E 快速的服務團隊 F 快速到府的服務團隊 G 快速寄送團隊。

A：發想…… B：目標…… C：願景……

A125: The company needs to run well. Must meet the following conditions. Ahigh-quality products B research and development product team C marketing team D fast maintenance team E fast service team F fast service team G fast delivery team.

A: Think... B: Objective... C: Vision...

A126：一個人對待每天，A 要有 100%的創意 B100%的努力 C100%渴望明天會更好更耀眼。

A：發想…… B：目標…… C：願景……

A126: One person treats every day. Amust have 100% creativity B100% effort C100% longing for tomorrow to be better and dazzling.
A: Think... B: Objective... C: Vision...

A127：夢想+時間表=目標　，方向+時間表=願景。
A：發想……　B：目標……　C：願景……
A127: Dream + timetable = goal, Direction + timetable = vision.
A: Think... B: Objective... C: Vision...

A128：微笑+熱情+關心=好人緣，不隨意指責人家+不隨意批評人家=好人緣　，多鼓勵人家+多讚美人家=好人緣。
A：發想……　B：目標……　C：願景……
A128: Smile + enthusiasm + care = good popularity. Indiscriminately accuse others + indiscriminately criticize others = good popularity Encourage others + praise others = good popularity.
A: Think... B: Objective... C: Vision...

A129：怎樣這一個全方面受歡迎的人，傾聽+多分享=受歡迎的人，多協助人+多付出=受歡迎的人，多惜福+多感恩=受歡迎的人。
A：發想……　B：目標……　C：願景……
A129: How about this popular person. Listen + share more = popular people. More helpers + more pay = popular people. More pity + more gratitude = popular people.
A: Think... B: Objective... C: Vision...

A130：別人不願意想的生意，事業，我們來想來做，發展空間就很大，機會空間就很深很廣很大。
A：發想……　B：目標……　C：願景……
A130: Business that others do not want to think about. cause. Let's

好的溝通能力+好的人際關係，是事業成功的關鍵所在。
Good communication skills + good interpersonal relationships. It is the key to successful career.

think about it. There is a lot of room for development. The opportunity space is very deep, wide and large.
A: Think... B: Objective... C: Vision...

A131：逆向思考模式……一次失敗，不代表 99%次都失敗，成功細節就在 99%次裡面。
A：發想…… B：目標…… C：願景……
A131: Reverse thinking mode...Afailure. It does not mean that it fails 99% of the time. The details of success are 99% of the time.
A: Think... B: Objective... C: Vision...

A132：A 態度決定命運，B 溝通決定命運，C 性格決定命運，要常慎思，深思。
A：發想…… B：目標…… C：願景……
A132: A attitude determines fate. B communication determines fate. C character determines fate. Always think carefully. Think deeply.
A: Think... B: Objective... C: Vision...

A133：壓力大，力量就大，知識制窮，知識致富，做事要細心・成就在專心。
A：發想…… B：目標…… C：願景……
A133: High pressure. The power is great. Poor knowledge. Knowledge becomes rich. Be careful in doing things. Achievement is in my heart.
A: Think... B: Objective... C: Vision...

A134：做對事，做好事，對的事放膽去做，大膽走出第一步，成功就在望，堅定信念，是成功的要件。
A：發想…… B：目標…… C：願景……
A134: Do the right thing. Do good things. Do the right thing boldly. Take the first step boldly. Success is in sight. Firm conviction. It is the

怎麼樣去發掘和開發自己潛在的能量，相信自己，肯定自己，自己是可以的，這是潛能開發最大的推動力。
How to discover and develop your potential energy. Believe in yourself, affirm yourself, and be yourself. This is the biggest driving force for potential development.

key to success.
A: Think... B: Objective... C: Vision...

A135：發想愈大，空間愈大，發想愈大，機會愈大，激盪腦力，創意無限。
A：發想…… B：目標…… C：願景……
A135: The more I think about it. The bigger the space. Think bigger. The greater the chance. Brainstorming. endless creativity.
A: Think... B: Objective... C: Vision...

A136：相信自己，就會成就自己，如果你想要，你就會擁有，所以成功，就是因為努力。
A：發想…… B：目標…… C：願景……
A136: Believe in yourself. Will achieve themselves. If you want. You will have. So success. It is because of hard work.
A: Think... B: Objective... C: Vision...

A137：除非你想富貴，否則富貴不會跟你，逐利概念，創造財富，永無壓力揚揚／剔？不高，人無壓力不成長，尋找轉機，創造巔峯。
A：發想…… B：目標…… C：願景……
A137: Unless you want to be rich. Otherwise the rich will not follow you. Profit-seeking concept. create wealth. Never stress is not high. People do not grow without pressure. Looking for a turnaround. Create the pinnacle.
A: Think... B: Objective... C: Vision...

A138：活到老學到老，活到老戰鬥老，戰鬥到最後一分一秒，一個人的成就，來自於不放棄，就會有所成就。

常常給予真誠的讚美和感謝，這是友誼和人際關係昇華的必備元素。
Always give sincere praise and gratitude. This is an essential element for the sublimation of friendship and interpersonal relationships.

A：發想…… B：目標…… C：願景……

A138: Too old to learn. Live to old warfare. Fight to the last minute and one second. Aperson's achievements. From not giving up. There will be achievements.

A: Think... B: Objective... C: Vision...

A139：模實紮實職場，堅持精緻任何事，著眼專業一點，讓專業事業占第一，新的時代，新的思維。

A：發想…… B：目標…… C：願景……

A139: Solid and solid workplace. Persevere in anything exquisite. Focus on professionalism. Let professional career take the first place. New era. New thinking.

A: Think... B: Objective... C: Vision...

A140：競爭的時代，新的創意，熟悉地理環境，有助業務事業發展，非凡事業成就。

A：發想…… B：目標…… C：願景……

A140: The era of competition. New ideas. Familiar with the geographical environment. Help business development. Extraordinary career achievements.

A: Think... B: Objective... C: Vision...

A141：面對客戶姿態對，客戶就會喜歡你，人生先有苦，才會有甘，一個客戶失了信，百客就不登門，重道德，守承諾，是做人，做事業的成功要件。

A：發想…… B：目標…… C：願景……

A141: Face the customer's attitude right. Customers will like you. Life is suffering first. Will be sweet. Acustomer has lost faith. Hundreds of guests will not go to the door. Focus on morality. Shou Chengnuo. Is a man. The key to success in your career.

A: Think... B: Objective... C: Vision...

常常引發他人心中的渴望，傾聽他人心中的需求，而給予充分的協助和幫助。

Frequently arouse the desires of others. Listen to the needs of others and give them full assistance and help.

A142：會讓利，才有可能得大利，小市場，小顧客，是創造企業版圖的種籽，別人賣有我賣缺，人家不做我去做，待人真誠，才會贏得客戶的信賴，誠懇的態度，熱情的語言，是經營致富的元素。
A：發想…… B：目標…… C：願景……
A142: Will benefit. Only then can it be profitable. Small market. Small customers. It is the seed for creating an enterprise layout. Someone else sells what I lack. If people don't do it, I will do it. Treat people sincerely. Will win the trust of customers. Sincere attitude. Passionate language. It is the element of business prosperity.
A: Think... B: Objective... C: Vision...

A143：業務的成敗，在於人與人之情感，而非金錢的條件，人是感性的動物，當你用真情打動它，他也會用真情回報你，台上一點鍾，台下十年功，持續，磨練。
A：發想…… B：目標…… C：願景……
A143: The success or failure of the business. It lies in human emotions. Not a condition of money. People are emotional animals. When you touch it with true feelings. He will also reward you with true feelings. One o'clock on the stage. Ten years off the stage. continued. Hone.
A: Think... B: Objective... C: Vision...

A144：碰到問題，困難狀況，就是去面對它，處理它，解決它，對事情才有助益，這就是人生哲學，發想是生意，事業的啟源路，開源路，切記。
A：發想…… B：目標…… C：願景……
A144: Encountered a problem. Difficult situation. Just face it. Deal with it. fix it. Only help things. This is the philosophy of life. Thinking is business. Qiyuan Road of Career. Open source road. Remember.
A: Think... B: Objective... C: Vision...

A145：發想要加以行動，執行，於是就能輝煌成功，玉不琢不成器，人不學不知義，己所不欲，勿施於人。

A：發想…… B：目標…… C：願景……

A145: I want to take action. carried out. So you can succeed brilliantly. Jade is not cut, not a device. People don't learn without understanding. Do what you want. Don't impose on others.

A: Think... B: Objective... C: Vision...

A146：帶人要帶心，賞罰要分明，內舉不避親，外舉不避仇，願無伐善，無施勞，心正，修身，齊家，治國，平天下。

A：發想…… B：目標…… C：願景……

A146: Take people with heart. Rewards and penalties should be clear. Don't avoid relatives within. Don't avoid enmity by extroverting. May nothing be good. No work. Heart is right. Slim. Qi family. Run the country. Flat world.

A: Think... B: Objective... C: Vision...

A147：健康三要素，睡眠，營養，運動，人無遠慮，必有近憂，目光要長遠，志向要遠大。

A：發想…… B：目標…… C：願景……

A147: Three elements of health. Sleep. nutrition. motion. There is no farsightedness. There must be near worries. Look long-term. The ambition is to be ambitious.

A: Think... B: Objective... C: Vision...

A148：勝不驕，敗不餒，創新和研發是致富的要素，能掌握人群的向心力，是成功的要件。

A：發想…… B：目標…… C：願景……

A148: Won't be proud. Undefeated. Innovation and R & D are the elements of getting rich. Can master the centripetal force of the crowd. It is the key to success.

A: Think... B: Objective... C: Vision...

A149：賭博和毒品，對人體健康，財產傷害最大，家庭和諧，是促成社會安定的元素。

A：發想…… B：目標…… C：願景……

A149: Gambling and drugs. For human health. Property damage is greatest. Family Harmony. It is the element that contributes to social stability.

A: Think... B: Objective... C: Vision...

A150：知識就是力量，求人不如求己，飲水要思源，知恩要報本，路遙知馬力，事久見人心，人為善，禍漸遠，福漸近。

A：發想…… B：目標…… C：願景……

A150: Knowledge is power. Asking for others is better than asking for yourself. Drinking water should be the source. Knowing gratitude to repay. Lu Yao knows horsepower. Seeing people for a long time. People are good. The disaster is far away. Blessing is approaching.

A: Think... B: Objective... C: Vision...

A151：一家之計在於和，一生之計在於勤，責人之心責己，恕己之心恕人，積善成名，積惡滅身。

A：發想…… B：目標…… C：願景……

A151: The family's plan lies in harmony. The whole life lies in diligence. Responsible people blame themselves. Forgive yourself. Jason became famous. Accumulate and die.

A: Think... B: Objective... C: Vision...

A152：遠水不救近火，遠親不如近鄰，有田不耕倉廩虛，有書不讀子孫愚，欲求生富貴，須下足功夫。

A：發想…… B：目標…… C：願景……

A152: Far water does not save near fire. Distant relatives are not as good as neighbors. Arables are not plowing, but they are empty. Don't read books for children and grandchildren. Want to live a prosperous life. It takes a lot of work.
A: Think... B: Objective... C: Vision...

A153：不經一事，不長一智，樂觀進取的態度，是生意事業成功的指標，成長需要毅力，生命需要韌性，知識+智識賺錢快，勞動賺錢慢，好好深思。
A：發想…… B：目標…… C：願景……
A153: Nothing happened. Not a wise one. Optimistic and aggressive attitude. Is an indicator of business success. Growth requires perseverance. Life needs toughness. Knowledge + wisdom makes money fast. Labor makes money slowly. Think deeply.
A: Think... B: Objective... C: Vision...

A154：我要做並展開行動，肯定自己，創造未來，我要試試，我想試試，是成功致富的開端，天時，地利，人和，三合一就所向無敵。
A：發想…… B：目標…… C：願景……
A154: I want to do it and take action. Affirm yourself. Create the future. I want to try. I want to try. Is the beginning of success. Days. Geographically. Renhe. Three in one is invincible.
A: Think... B: Objective... C: Vision...

A155：一個人的思維，一個人的意志力，大大改變一個人的命運，常生氣，疾病會越來越多，常知足，快樂會越來越多。
A：發想…… B：目標…… C：願景……
A155: One's thinking. One's willpower. Greatly change the fate of a person. Always angry. There will be more and more diseases. Often contented. Happiness will increase.

愛茵斯坦說，一個人的成功常取決於轉折點上。
Einstein said that one′s success often depends on the turning point.

A: Think... B: Objective... C: Vision...

A156：常抱怨，煩惱會越來越多，喜歡占便宜，會越來越貧窮，喜歡享福，會越來越痛苦，喜歡學習，會越來越有智慧。
A：發想…… B：目標…… C：願景……
A156: I often complain. Trouble will increase. Like to take advantage. Will become more and more poor. Like to enjoy. Will be more and more painful. Like learning. Will become more and more wise.
A: Think... B: Objective... C: Vision...

A157：一個人的成功，並不完全取決起跑上·而大部分取決於轉捩上，或挫折點上，因挫折激發你潛在常沒有發揮的能力，沒有發掘的潛在能力，挫折是機會來的開端，適時抓住機會，掌握機會，這是成功不可或缺的重要元素。
A：發想…… B：目標…… C：願景……
A157: One's success. It does not entirely depend on the start. And most depends on the transition. On the setback point. Frustration motivates you to potentially fail to perform. No perceived potential. Frustration is the beginning of opportunities. Seize the opportunity at the right time. Take advantage of opportunities. This is an indispensable and important element of success.
A: Think... B: Objective... C: Vision...

158：人生的價值觀……就是做自己最有兴趣事，也是有能力做的與人分享，這就是人生的價值觀。
A：發想…… B：目標…… C：願景……
A158: The values of life... is to do what you are most interested in. It is also capable of sharing with others. This is the value of life.
A: Think... B: Objective... C: Vision...

A159：隨時隨地，要充實自己自信，溝通能力，抗壓力，要充實自己的內涵和實力，只要勤奮努力，成功就會離你愈來愈近，這是自然法則。

A：發想…… B：目標…… C：願景……

A159: Anytime, anywhere. To enhance their self-confidence. Communication skills. Anti-stress. To enrich their connotation and strength. Just work hard. Success will be closer and closer to you. This is the law of nature.

A: Think... B: Objective... C: Vision...

A160：用熱情的服務的心，用品質顧好的心，體貼的心，來擄獲顧客的心。

A：發想…… B：目標…… C：願景……

A160: With a heart of warm service. Take good care of quality. Thoughtful heart. Come to capture the hearts of customers.

A: Think... B: Objective... C: Vision...

A161：強化輔助推銷，強化幫助推銷，強化協助推銷。

A：發想…… B：目標…… C：願景……

A161: Strengthen auxiliary sales promotion. Strengthen help sales. Strengthen assist sales.

A: Think... B: Objective... C: Vision...

A162：常逃避，失敗越來越多，常努力，成功越來越多，常付出，福報越來越多，常助人，貴人越來越多。

A：發想…… B：目標…… C：願景……

A162: Escape often. More and more failures. Always work hard. More and more success. Always pay. More and more blessings. Often helping others. More and more nobles.

A: Think... B: Objective... C: Vision...

隨時檢視自己的工作效率，想辦法倍增工作效率。

Check your work efficiency at any time. Find ways to increase work efficiency.

A163：常分享，朋友越來越多，常感恩，順利越來越多，常施財，富貴越來越多。

A：發想…… B：目標…… C：願景……

A163: Share often. More and more friends. Always grateful. More and more smoothly. Chang Shicai. More and more wealthy.

A: Think... B: Objective... C: Vision...

A164：訓練不只是指技巧，而是精神，公司人才愈多，營利就愈高，公司服務愈好，收入就愈好。

A：發想…… B：目標…… C：願景……

A164: Training is not just about skills. It is spirit. The company has more talents. The higher the profitability. The better the company's services. The better the income.

A: Think... B: Objective... C: Vision...

A165：明確的目標，早晚大聲朗誦……鋼鐵大王卡耐基，充分利用潛意識，徹底思考每一個問題……發明大王愛迪生，不浪費時間在無益的上面，要有強烈的企圖心……汽車大王亨利福特。

A：發想…… B：目標…… C：願景……

A165: Clear goals. Speaking aloud sooner or later... Carnegie the Steel King. Make full use of the subconscious. Think about every problem thoroughly... Inventor Edison. Do not waste time on the unprofitable. There must be a strong ambition... Automobile King Henry Ford.

A: Think... B: Objective... C: Vision...

A166：運用想像力，先行演練，事先計劃……油輪大亨歐納西斯，懂得如何激勵自己和屬下……克萊斯勒汽車艾柯卡，勇氣比知識更重要……受重傷N次的巨星成龍。

A：發想…… B：目標…… C：願景……

銷售之所以成功，就是能掌握市場的需求，大量的行銷概念。

The reason why sales are successful is to be able to grasp the needs of the market. A large number of marketing concepts.

A166: Use your imagination. Advance performance. Plan ahead... Tanker tycoon Onassis. Know how to motivate yourself and your subordinates... Chrysler Auto Ecoca Courage is more important than knowledge... The superstar Jackie Chan, who was seriously injured N times.
A: Think... B: Objective... C: Vision...

A167：一分耕耘，一分收穫，種瓜得瓜，種豆得豆，銷售本質就是不停的推銷，成功機率就大，人生就是要翻轉，翻轉人生才有希望，投資自己的未來，做好準備補充動能，如知識技能和經驗。
A：發想…… B：目標…… C：願景……
A167: Hard work. One point harvest. Growing melons. Plant beans to get beans. The essence of sales is non-stop marketing. The chance of success is greater. Life is about turning. There is hope in turning over life. Invest in your own future. Be prepared to add kinetic energy. Such as knowledge skills and experience.
A: Think... B: Objective... C: Vision...

A168：做事情，做比說重要……知難行易，服務好，評價高，來自於你的細心，用心，真心，改變思維，才能改變你的前途，創造事業巔峯。
A：發想…… B：目標…… C：願景……
A168: Do things. It is more important to say than to know... good service. High evaluation. From your carefulness. Attentively. sincere. Change thinking. In order to change your future. Create a career peak.
A: Think... B: Objective... C: Vision...

A169：用強大的信念改造自己，用強大的信念創造財富，人要善於理財，懂得生活，生活才會穩定'幸福，美滿，原有的客戶很重要，但不斷開發新客戶更重要。

隨時掌握最新的資訊，隨時檢討，分析，做決策，納入公司決策一環，並確實實行之。

Keep up to date with the latest information at any time. Review, analyze, make decisions at any time, and incorporate it into the company's decision-making process.

A：發想…… B：目標…… C：願景……

A169: Transform yourself with strong faith. Create wealth with strong beliefs. People should be good at financial management. Understand life. Life will be stable 'happiness. happy. The original customer is very important. But constantly developing new customers is more important.

A: Think... B: Objective... C: Vision...

A170：朋友相處要互相支持，鼓勵，幫助，人際關係愈好，路愈來愈廣，銷售業務的決勝關鍵點，在於做人，做事，社交圓融，讓客戶100%相信你，你終必得到一大筆生意，把每一件事情做到好，做到最極致，那你就成功了。

A：發想…… B：目標…… C：願景……

A170: Friends should support each other when they get along. encourage. help. The better the interpersonal relationship. The road is getting wider and wider. The key to winning the sales business. Lies in life. work. Social harmony. Let customers believe you 100%. You will eventually get a lot of business. Do everything well. Do the most extreme. Then you succeeded.

A: Think... B: Objective... C: Vision...

A171：無友不如己者，過則勿憚改，三思而後行，是非只為多開口，煩惱皆因強出頭，忍一時之氣，免百日之憂。

A：發想…… B：目標…… C：願景……

A: 171. No friend is worse than yourself. Don't worry about changing. Think twice. Is it just for multiple openings. Troubles are all because of the strong. Endure for a while. No worries for a hundred days.

A: Think... B: Objective... C: Vision...

A172：人無橫財不富，馬無夜草不肥，人惡人怕天不怕，人善人欺天不欺，得寵思辱，居安思危。

隨時分析自己的優點和缺點，隨時擴大優點，縮減缺點，使自己變得更優質，更卓越，更精進。

Analyze your own strengths and weaknesses at any time. Expand your strengths at any time. Reduce your weaknesses. Make yourself better, better, more sophisticated.

A：發想…… B：目標…… C：願景……

A172: People are not rich without wealth. Ma Wuye grass is not fat. Everyone is afraid of the sky or not. Good people do not bully the sky. Be petty and shameful. Stay in peace.

A: Think... B: Objective... C: Vision...

A173：少壯不努力，老大徒傷悲，天網恢恢，疏而不漏，從儉入奢易，從奢入儉難。

A：發想…… B：目標…… C：願景……

A173: Young and strong do not work hard. The boss is sad. Skynet was restored. Sparse but not leaking. From frugality to extravagance. From extravagance to frugality.

A: Think... B: Objective... C: Vision...

A174：路不行不到，事不為不成，忠言逆耳利於行，良藥苦口利於病，滾石不生苔，前進，奮鬥.向成功之路大步邁進。

A：發想…… B：目標…… C：願景……

A174: The road can't be reached. Nothing is impossible. Loyalty is conducive to action. Good medicine is good for bitterness. Rolling stones do not produce moss. go ahead. Struggle. Stride towards the road to success.

A: Think... B: Objective... C: Vision...

A175：創意思考讓成就加分，積極努力放胆去闖，成功者很少抱怨。

A：發想…… B：目標…… C：願景……

A175: Creative thinking adds points to achievement. Actively work hard to go boldly. Successful people rarely complain.

A: Think... B: Objective... C: Vision...

A176：成功者善用時間，成功者尋找方法，成功者懂得謙虛，成功者洞燭機先。

A：發想…… B：目標…… C：願景……

A176: Successful people make good use of time. Winners find ways. The winner understands modesty. The winner hole candle machine first.

A: Think... B: Objective... C: Vision...

A177：成功者創意思維，成功者活用人際，成功者肯定自信，自我，成功者樂觀進取。

A：發想…… B：目標…… C：願景……

A177: Creative thinking of successful people. Successful use of interpersonal. The winner must be confident. self. The winners are optimistic and enterprising.

A: Think... B: Objective... C: Vision...

A178：井易免店面飲水機，淨水設備加盟事業，讓你學會知識（Know-H0W），與技術，開發潛能，開發做不完的客戶，收入持續增加，當副業，創業最好的選擇。

A：發想…… B：目標…… C：願景……

A178: Jingyi free store water dispenser. Join the watcr purification equipment business. Let you learn knowledge（Know-H0W）. And technology. Dcvelopment potential. The development of the endless customer service. Income continues to increase. When sideline. The best option for starting a business.

A: Think... B: Objective... C: Vision...

A179：給客戶最高的滿意度，業績會蒸蒸日上，不要蹉跎歲月，無所事事，你的一生靠你自已主宰。

A：發想…… B：目標…… C：願景……

隨時強化專業技能，隨時強化人際溝通能力，這是成功的養分，也是成功的基石。

Strengthen professional skills at any time. Strengthen interpersonal communication skills at any time. This is the nutrient of success. It is also the cornerstone of success.

A179: Give customers the highest satisfaction. Performance will be booming. Don't waste the moon. Do nothing Your life depends on you.
A: Think... B: Objective... C: Vision...

A180：錢不是萬能自己的價值，沒有志向，就像船沒航向，迷迷茫茫，不會創造輝煌。
A：發想…… B：目標…… C：願景……
A180: Money is not a universal value. No ambition. It's like the ship is not heading. Confused. Will not create brilliance.
A: Think... B: Objective... C: Vision...

A181：不批評，不責備，不抱怨，是好人緣，好人際的開端。
A：發想…… B：目標…… C：願景……
A181: No criticism. No blame. Do not complain. Is a good relationship. The beginning of a good relationship.
A: Think... B: Objective... C: Vision...

A182：一枝草，一點露，只要奮鬥，就會出路。
A：發想…… B：目標…… C：願景……
A182: A piece of grass. Alittle exposed. As long as Fen Yu. There will be a way out.
A: Think... B: Objective... C: Vision...

A183：一分努力，一分收穫，走對的路。
A：發想…… B：目標…… C：願景……
A183: A little effort. One point harvest. Go the right way.
A: Think... B: Objective... C: Vision...

A184：用行動來激勵自己，讓自己邁向有希望的未來。
A：發想…… B：目標…… C：願景……
A184: Motivate yourself with actions. Let yourself move towards a promising future.
A: Think... B: Objective... C: Vision...

A185：做人，做事，做事業，適度的公共關係，有助於事業的整體發展。
A：發想…… B：目標…… C：願景……
A185: Life. work. Do a career. Moderate public relations. Contribute to the overall development of the cause.
A: Think... B: Objective... C: Vision...

A186：合人，合群，合諧=三合，是創造多贏的人際關係。
A：發想…… B：目標…… C：願景……
A186: Together. Group. Harmony = Sanhe. It is to create win-win relationships.
A: Think... B: Objective... C: Vision...

A187：一個人的思維，一個人的意志，將大大改變一個人的命運。
A：發想…… B：目標…… C：願景……
A187: One's thinking. One's will. Will greatly change one's fate
A: Think... B: Objective... C: Vision...

A188：成功者不吝付出，成功者善頌善禱，成功者注意形象。
A：發想…… B：目標…… C：願景……
A188: Successful people don't hesitate to pay. The winners sing praises and prayers. Winners pay attention to the image.
A: Think... B: Objective... C: Vision...

成功旅途上是艱辛的，是辛苦的，是充實的，是快樂的，絕不退縮，絕不放棄，這是成功唯一的心態和信念。
On the journey of success. It is hard. It is hard. It is full. It is happy. Never back down. Never give up.

A189：成功者以德服人，成功者志向遠大，成功者全力以赴，成功者多行善事。

A：發想…… B：目標…… C：願景……

A189: Winners serve people with virtue. The winners have great ambitions. Winners go all out. Winners do more good deeds.

A: Think... B: Objective... C: Vision...

A190：想別人不想做的事業，做別人不想做的事業，創造別人不想做的事業，這樣必能有所發展，有所成就。

A：發想…… B：目標…… C：願景……

A190: Think about what others do not want to do. Do what others don't want to do. Create a cause that others do not want to do. This will certainly develop. Achievement.

A: Think... B: Objective... C: Vision...

A191：知識力+想像力+行動力，是致富的成功的要件。

A：發想…… B：目標…… C：願景……

A191: Knowledge + imagination + action. It is the key to success.

A: Think... B: Objective... C: Vision...

A192：一百個想法，不如一個行動。

A：發想…… B：目標…… C：願景……

A192: One hundred ideas. Better than an action.

A: Think... B: Objective... C: Vision...

A193：做生意，做事業，務必一步一腳印，在穩定中求發展，在穩定中求壯大，這樣方可立於不敗之地。

A：發想…… B：目標…… C：願景……

A193: Doing business. Do a career. Be sure to step by step. Seek

development in the setting. Strive for growth in stability. Only in this way will it be invincible.
A: Think... B: Objective... C: Vision...

A194：一個技術，一個工具袋，一個產品知識，改變你一生的命運。
A：發想…… B：目標…… C：願景……
A194: A technology. Atool bag. Aproduct knowledge. Change the destiny of your life.
A: Think... B: Objective... C: Vision...

A195：馬不吃夜草不肥，人不努力就不長進，就不會有所發展，有所成就。
A：發想…… B：目標…… C：願景……
A195: Horses do not eat wild grass and are not fat. People do not grow without effort. There will be no development. Achievement.
A: Think... B: Objective... C: Vision...

A196：多學習，多觀摩，多研習，多投資.投資自己是穩賺不賠的。
A：發想…… B：目標…… C：願景……
A196: Learn more. Observe more. Learn more. Invest morc. Investing yourself is profitable.
A: Think... B: Objective... C: Vision...

A197：沒有推銷不出去的產品，只有自己不想要把產品推銷出去。
A：發想…… B：目標…… C：願景……
A197: There are no products that cannot be sold. Only I don't want to sell the product.

A: Think... B: Objective... C: Vision...

A198：前進的路，推廣的路，戰鬥的路，永不畏懼，永不放棄。

A：發想…… B：目標…… C：願景……

A198: The way forward. Road to promotion. The road to warfare. Never be afraid. never give up.

A: Think... B: Objective... C: Vision...

A199：銷售的技巧，就是概念與態度的改變，其結果就會不一樣。

A：發想…… B：目標…… C：願景……

A199: Sales skills. Is the change of concept and attitude. The result will be different.

A: Think... B: Objective... C: Vision...

A200：改變窮腦袋，才有富口袋。

A：發想…… B：目標…… C：願景……

A200: Change the poor head. Only have rich pockets.

A: Think... B: Objective... C: Vision...

A201：成功人士的看法不只是要成功，或希望成功，而是強烈的渴望成功。

A201: The view of successful people is not just to succeed, or to hope for success, but a strong desire for success.

A202：一個小孩看到喜歡的玩具，先告訴媽媽，媽媽不答應就找爸爸懇求，爸爸也不答應就哭起來，甚至坐在地上兩腳亂

成功是靠有準備的，成功是靠有機運的，成功是靠有責任心的，成功是靠有實力的。

Success depends on preparation. Success depends on luck. Success depends on responsibility. Success depends on strength.

踢、這是原始的為成功找方法。

A202: When a child sees his favorite toy, he first tells his mother that if he does not agree, he will ask his father to plead, and if he does not agree, he will cry, or even sit on the ground and kick his feet. This is the original way to find success.

A203：選擇比努力更重要,要到達目的地、騎自行車 1 小時 10 多公里,汽車可以跑 100 公里、坐上飛機邊吃美食居然可以跑 1000 公里、平台不同而已。

A203: Choice is more important than hard work. To reach the destination, to ride a bicycle for more than 1 hour and 10 kilometers, the car can run for 100 kilometers, and eat on the plane while eating food can actually run for 1,000 kilometers.

A204：找對的路徑,結交好朋友成為你的貴人,有正能量的,肯指點你的朋友（因為當局著迷、不知自己的缺點）領你走過泥濘,當你的墊腳石的貴人。

A204: Find the right path, make good friends and become your noble person, with positive energy, willing to point your friends（because of the fascination of the authorities and unaware of your own shortcomings）lead you through the muddy, as your stepping stone Noble.

A205：有領導能力的人,經常是會向群眾導引想法及做法的人。

A205: People with leadership skills are often people who can guide ideas and practices to the masses.

A206：遙遠的理想目標有些模糊不清，「做好正在做的事情」這是我的老師葉世傳先生的座右銘,也是我遵行的準則。

A206: The distant ideal goal is a little vague. "Do what you are doing."

窮則變，轉彎，變則通。

The poor is the way. The turn. The way is the way.

This is the motto of my teacher, Mr. Ye Shichuan. It is also the principle that I follow.

A207：我的校長楊展雲先生勉勵我們讀書、努力讀書、工作、認真工作、遊玩、盡情的玩。

A207: My principal Mr. Yang Zhanyun encouraged us reading, studying hard, working, working seriously, playing and playing.

A208：為明天的再出發做準備，最好的做法就是把所有腦力、精力、熱忱完全發輝，把握當下。

A208: To prepare for tomorrow's re-start, the best way is to fully radiate all your brain power, energy, and enthusiasm to grasp the moment.

A209：一位老爸要測試三個兒子，看誰最會做生意，拿出了市價等值的四個蘋果、一個大西瓜、十個檸檬看誰能賣出最好的價錢，大哥選了蘋果，二哥選了大西瓜，小弟只好選檸檬，小弟把檸檬榨汁加上愛玉加上糖水，而賣出多幾倍的價錢，這就是有動腦的結果。

A209: A dad wants to test three sons to see who can do business the most, and comes up with four apples, one big watermelon and ten lemons at market value to see who can sell the best price. The eldest brother chose the apple, the second elder chose the big watermelon, the younger brother had to choose the lemon, the younger brother squeezed the lemon juice plus Aiyu plus sugar water, and sold several times the price.

A210：新開發出一種利用離心力脫水的拖把，用腳踩旋轉，雙手不用碰水是一種很好的商品，後來有人研發出用手壓拖把也可以旋轉脫水，是另一種不同型式的開發，身邊有那些可以改

走過的路，走過的經驗，就是你的資產。

The road traveled. The experience traveled is your asset.

進的，動動腦會有所收獲。
A210: Newly developed a mop that uses centrifugal force for dehydration. It is a good product to rotate with your feet and do not need to touch the water with your hands. Later, someone developed a mop that can be rotated by hand. , There are those around you that can be improved, and your brain will gain something.

A211：幼童的教育首重啟發與導引，會動腦走向正途的觀念將影響他的一生，我的小孩讀幼兒園時，感冒發燒去醫院必須打針，我告訴他因為你發燒一定要打針才會好，而打針像蚊子咬一點點痛而已，結果我兒子勇敢的面對打針而不哭，同一時間有一位和他同年紀的男童也須打針，却大聲哭喊,兩個大人抓住還不停的扭動，我很担心針頭會不會斷在肌肉裡，這就是有沒有和幼童溝通而已。
A211: The education and guidance of young children's education will be the first to restart, and the concept of moving the brain to the right way will affect his life.When my child is in kindergarten, he must have an injection when going to the hospital with a cold and fever. I tell him because you must have a fever. The injection will be good, and the injection is like a mosquito biting a little bit of pain.As a result, my son bravely faced the injection without crying.At the same time, a boy of the same age as him also had to give the injection, but he cried out loud, two The adult grasped and kept twisting, and I was worried that the needle would break in the muscle. This is whether I communicated with the young child.

A212：我的小兒子 4 歲時沒有用工具把我的防風打火機細部分解，我看到以後不但沒責罵他，反而稱讚他是拆船公司董事長，我的機械概念很強，都感到不容易這麼拆卸，果然印證他長大以後，機械概念比我強，這是稱讚和鼓勵的結果。
A212: When my younger son was 4 years old, he didn't use tools to dissolve my windproof lighter in detail. I saw that instead of scolding

him, I praised him as the chairman of the ship recycling company. My mechanical concept is very strong. I feel that it is not easy to disassemble this way, and it turns out that after he grew up, the mechanical concept is stronger than mine, which is the result of praise and encouragement.

A213：小孩子有些是很頑皮的過動兒，可以用導引他的好奇和精力的舒發，不可一直用責罵和禁止，我用枕頭讓精力過剩的兒子練拳擊，消耗他的體力，當時湯瑪士，愛迪生把母雞趕走自己去的孵蛋，如果父母禁止的話就沒有（發明大王）的封號了。

A213: Some children are naughty and hyperactive. You can use the curiosity and energy to guide him. Do n't use scolding and prohibition all the time. I use pillows to let my son with excess energy practice boxing Physical strength, then Thomas. Edison drove the hen away from the hatching eggs, and if the parents prohibited it, there would be no title（the king of invention）.

A214：如果你會給你的小孩愈挫愈勇的信念，給他積極正向的思考，而且有決心和毅力去完成，他的成就會展現在你的面前。

A214: If you will give your child more and more conviction, give him positive and positive thinking, and have the determination and perseverance to complete, his achievement will be displayed in front of you.

A215：如果小孩有語言或動作等方面的缺失，現在已經有復健方面的技術可以改良和糾正，要走正確的管道才會有好的結果。

A215: If the child has a lack of language or movement, there are already rehabilitation techniques that can be improved and corrected.

Only by taking the correct pipeline will there be good results.

A216：美國保險業曾測試和調查 20 歲的高中畢業生 100 名，問他們將來要成功否？有 100 人都舉手,經過 45 年後調查，其中成功者 5 位，仍然在上班的 5 位，死亡 36 人，一文不名須靠外力協助的有 54 人，這是大自然生態的數字。

A216: The US insurance industry has tested and surveyed 100 20-year-old high school graduates and asked them whether they will succeed in the future? 100 people raised their hands. After 45 years, 5 of them were investigated, and 5 still At work, 36 people died, and 54 people needed external assistance. This is the number of natural ecology.

A217：首先探討失敗的原因：1.對自我不忠實：得過且過。2.藉口特別多：沒時間……等等。3.沒有目標：如同是沒有舵的船。4.態度不好的人：內在、外在。5.沒有堅持：沒有下定決心持續的努力。

A217: First discuss the reasons for failure, 1. Unfaithfulness to self: to live and live. 2. There are many excuses: no time... wait etc. 3: No goal: like a ship without a rudder. 4: People with bad attitudes: internal and external. 5: No persistence: No determination to continue efforts.

A218：黃金成功定律：1.明確的目標，數字或位階明確。2.有計劃的文書化：如何去達成？肯付出多少心力？3.設定期限：完成的期限短、中、長期。4.立即行動：動用腦力持續努力與堅持。5.早上晚上大聲朗誦：讓你的渴望，浮現眼前相信自己可以達成。

A218: The golden law of success: 1. Clear objectives. The number or rank is clear. 2. Planned paperwork: how to achieve it? How much effort is willing to pay? 3. Set the deadline: the deadline is short, medium and long. 4. Act immediately: use your brain to continue your efforts and persistence. 5. Read aloud in the morning and evening: let your longing come forward and believe that you can achieve it.

A219：萊特兄弟夢想讓人類可以像鳥一樣在天空飛翔，因為有他們的堅持和努力付出，才有今日的超音速飛機、以及可以進入太空的太空梭。
A219: The Wright brothers dream of allowing humans to fly in the sky like a bird. Because of their persistence and hard work, only today's supersonic aircraft and space shuttle can enter space.

A220：美國鋼鐵大王安德魯・卡內基和台灣經營之神王永慶先生，教育程度不高，出身低微，但是都有明確的目標而去執行，一步一步的拓展事業。
A220: American steel king Andrew Carnegie and Taiwanese operating god Mr. Wang Yongqing have low education and low background, but they all have clear goals to implement and expand their business step by step.

A221：不要讓貧窮侵蝕了你的夢想，不要讓消極觀念，扼殺你的未來，只有相信自己可以做得到的人，才會擁有財富。
A221: Don't let poverty erode your dreams, don't let negative ideas stifle your future, only people who believe they can do it will have wealth.

A222：給他一大堆的魚，不如教他如何釣魚，這樣可以自立更生，取之不盡。
A222: Give him a lot of fish, it is better to teach him how to fish, so that he can be self-reliant and inexhaustible.

A223：信心是一切事情要成功的基本要件，信心來自潛意識，告訴自己可以達成，讓勇氣和信心驅使意念轉化為事實。
A223: Confidence is a basic requirement for everything to succeed. Confidence comes from the subconscious, telling yourself that it can be achieved, and letting courage and confidence drive ideas into reality.

A224：人的一生只有透過學習，才能找到方法，逐步改善自己得到快樂，得到財富。
A224: Only through study can one find a way to improve one's own life and get happiness and wealth.

A225：行進中不可能永遠不會跌倒，重要的是跌倒後再爬起來，得到經驗，不會犯同樣的錯誤而可以愈來愈順利。
A225: It is impossible to never fall while traveling. The important thing is to climb up after falling down to gain experience and not make the same mistakes but can get smoother and smoother.

A226：了解自己的缺點愈多，透過反省、檢討,成長的速度愈快，對自己的信心，也會愈堅定。
A226: The more you understand your shortcomings, the faster you grow through reflection and review, and the firmer your confidence in yourself will be.

A227：我們的日落是另一個半球別的國度的日出，人生會有燦爛與晦暗之時，熬過了黑暗終究會得到光明。
A227: Our sunset is the sunrise of another country in the other hemisphere. When life will be bright and dark, we will get light after the darkness.

失敗，挫折，打擊，羞辱，都是你再一次成長的養分。
Failure, frustration, thrashing, humiliation are all the nutrients you grow up again.

A228：路徒雖然很近，不走是不會到達的，事情雖然小，不去做是不會成功的。

A228: Although the Luthians are very close, they won't arrive if they don't walk. Although things are small, they won't succeed unless they do it.

A229：學習如何獲得新知識、新資訊，學習如何選擇，學習如何做好人際關係，是恒久不變的學習重點。

A229: Learning how to acquire new knowledge and information, learning how to choose, and how to make good interpersonal relationships are the constant learning priorities.

A230：理想是明燈，沒有明燈就沒有方向，理想口頭說是口說無憑，理想寫在紙上是紙上談兵，理想要付諸行動才能實現。

A230: The ideal is a bright light. Without a bright light, there is no direction.

A231：健康的三大要素是：一睡眠，二營養，三運動，三者須兼顧。

A231: The three major elements of health are: one sleep, two nutrition, three exercise, and the three must be taken into consideration.

A232：人生須要追求五大健康：一心理健康、二身體健康、三經濟健康、四家庭健康、五社會健康。

A232: There are five major health needs in life: one mental health, two physical health, three economic health, four family health, and five social health.

A233：老天爺給每一個人都是一天 24 小時，非常公平，其實時間像乳溝一樣硬擠就有了。
A233: God gives me a person 24 hours a day, which is very fair. In fact, the time is as hard as the cleavage.

A234：人比人會氣死人，要和自己賽跑,每天、每月、每年是否有所成長這才是最重要的。
A234: People are more angry than others, and they have to race with themselves. Whether it grows every day, every month or every year is the most important thing.

A235：一個事業體、甚至一個國家，必須有賢能的人做事，這個事業體才會獲利成長，這個國家才能富強安定。
A235: A business body, or even a country, must have talented people to do things, this business body will profitably grow, and this country can be strong and stable.

A236：人際關係要做得好，首重讚美與鼓勵，任何人都喜歡誠心的讚美,屬下的成就來自你的鼓勵，對你個人都有加分的作用。
A236: Interpersonal relationship needs to be done well. The first emphasis is on praise and encouragement. Everyone likes sincere praise. The achievements of your subordinates come from your encouragement and have a personal bonus point.

A237：會把握機會的人，就是創造時勢的英雄，加上行動，努力付出，加上堅持，成功在望 e。
A237: The person who can seize the opportunity is the hero who creates the situation, plus action, hard work, plus persistence, success is in sight e.

校長，企業家，領導人，就像花圃的園丁，天天灌溉每一顆種籽，天天照顧每一顆種籽（員工），快快成長，健康壯大。
Principal. Entrepreneur. Leader. Just like the gardener of the flowerbed. Irrigating every seed every day. Taking care of every seed every day（employee）. Growing fast. Healthy and strong.

A238：只要你肯開口，到處是機會，到處是人群……
A：發想…… B：目標…… C：願景……
A238: As long as you are willing to speak up. Opportunities are everywhere. Crowds everywhere...
A: Think... B: Objective... C: Vision...

A239：成功不在你跨出去的腳步大或小，而是在你的方向正不正確，這是成功關鍵所在.切記。
A：發想…… B：目標…… C：願景……
A239: Success is not in the size of the steps you take out. It is in the wrong direction. This is the key to success. Remember.
A: Think... B: Objective... C: Vision...

A240：在業務上，在事業上，在生意上，要讓客戶覺得有感動，在服務上，要讓客戶感覺有感動的氛圍，感恩的氛圍，這就是銷售成功的關鍵所在，切記。
A：發想…… B：目標…… C：願景……
A240: In the business. In the business. In the business. To make the customer feel moved. In the service. To the customer to feel the moved atmosphere. Thanksgiving atmosphere. This is the key to successful sales. Remember.
A: Think... B: Objective... C: Vision...

A241：一個領導者，一個成功者，總是全力以赴，投入和持續不斷的努力，打拼，而這是領導者，成功者，必然的走向，也是必然的結果.切記。
A：發想…… B：目標…… C：願景……
A241: A leader. Awinner. Always go all out. Devotion and continuous efforts. Hard work. And this is the leader. The winner. The inevitable trend. It is also the inevitable result. Remember.

只要肯上進，肯努力，貧窮也能變富有。
As long as you are willing to make progress, be willing to work hard, and poverty can become rich.

A: Think... B: Objective... C: Vision...

A242：推銷的力量就是讓消費者，有被愛的感覺，有愛的力量，有感動的力量，世界最強的兵器就是感動的兵器。
A：發想…… B：目標…… C：願景……
A242: The power of marketing. It is to give consumers the feeling of being loved. The power of love. The power of touch. The strongest weapon in the world. The weapon of touch.
A: Think... B: Objective... C: Vision...

A243：在生活中，在工作中，在任務中，在事業中，常常要告訴自己，我心中沒有難這個字，沒有不可能這些字，只有 YES 這個字.切記。
A：發想…… B：目標…… C：願景……
A243: In life. At work. In tasks. In career. I often have to tell myself. There is no word in my heart. There is no word in impossible. There is only the word YES.
A: Think... B: Objective... C: Vision...

A244：想……這個字很重要，因為有想，才有計劃，再付出行動，那麼成功是必然的，這就是築夢踏實，切記。
A：發想…… B：目標…… C：願景……
A244: Think... This word is very important. Because you have thought. Only have a plan. Then take action. Then success is inevitable. This is to build a dream.
A: Think... B: Objective... C: Vision...

A245：每一個人，每一天的每一步路，都在做決定，做選擇，生活的決定與選擇，對或錯，深深的影響著你未來的前途，所以慎思明斷，思而後行，是決定未來的前途所在，切記。
A：發想…… B：目標…… C：願景……
A245: Everyone. Every walk of every day. Make decisions. Make choices. Life decisions and choices. Right or wrong. Deeply affect your future. So think wisely. Think before you act. Yes Decide where the future lies. Remember.
A: Think... B: Objective... C: Vision...

A246：信賴度，是客戶擴散和外溢的培養土，因為信賴，所以客戶會介紹更多的客戶，這是不變的自然法則，切記。
A：發想…… B：目標…… C：願景……
A246: Reliability. It is the cultivation ground for the proliferation and spillover of customers. Because of trust, customers will introduce more customers. This is a natural law that is not bad. Remember.
A: Think... B: Objective... C: Vision...

A247：改變和突破瓶頸，改變求生存，是企業求發展，求壯大最重要的法寶，切記。
A：發想…… B：目標…… C：願景……
A247: Change and break through the bottleneck. Change for survival. It is the most important magic weapon for enterprises to seek development. Seek growth. Remember.
A: Think... B: Objective... C: Vision...

A248：一個最會銷售的人，必定是一個最會溝通的人，一個最成功的人，必定是一個最會溝通的人，一個最成功的領導人，也必定是一個最會溝通的人，共勉之。
A：發想…… B：目標…… C：願景……
A248: A person who can sell the most. Must be a person who can communicate the most. Aperson who can communicate the most. Aperson who can communicate the most. Aleader who can

communicate the most. Encouragement.
A: Think... B: Objective... C: Vision...

A249：相信自己，肯定自己，努力自己，奮鬥自己，勇往向前不懈，向目標前途大步邁進，大步向前努力，這是成功的所在，共勉之。
A：發想…… B：目標…… C：願景……
A249: Believe in yourself. Affirm yourself. Strive for yourself. Struggle for yourself. Go forward courageously and relentlessly. Encouragement.
A: Think... B: Objective... C: Vision...

A250：產品本質要好，自然銷售就好，重點不在客戶買或不買，而在於你要把產品怎麼來的故事告訴客戶，產品優點告訴客戶，要詳細傳達給客戶和瞭解，雖然暫時沒有買，當客戶跟朋友閒聊時，也會有意無意的幫你散播產品的好，因為散播的力量無限大，切記。
A：發想…… B：目標…… C：願景……
A250: The essence of the product is better. The natural sales is better. The focus is not on whether the customer buys or not. It lies in the story of how you want the product to come. Tell thc customer. Tell the customer about the advantages of the product. Did not buy it. When customers chat with friends, they will help you spread the product intentionally or unintentionally. Because the power of spreading is infinite. Remember.
A: Think... B: Objective... C: Vision...

A251：不斷的前進，持續的努力，是致富成功的唯一的路。
A：發想…… B：目標…… C：願景……
A251: Keep moving forward. Continuous efforts. Is the only way to get rich.

A: Think... B: Objective... C: Vision...

A252：常對自己說，我是最優秀的，我是最有活力的，我一定能做的最好。
A：發想…… B：目標…… C：願景……
A252: I often tell myself. I am the best. I am the most energetic. I can do the best.
A: Think... B: Objective... C: Vision...

A253：一個人的態度，將決定自己的命軍，一個人的性格將決定你的成敗。
A：發想…… B：目標…… C：願景……
A253: One's attitude. Will determine his own destiny。
Aperson's character will determine your success or failure.
A: Think... B: Objective... C: Vision...

A254：夢想，勇氣，韌性，戰鬥力是成功致富的要件。
A：發想…… B：目標…… C：願景……
A254: Dream. courage. The nature. Fighting power is the key to success.
A: Think... B: Objective... C: Vision...

A255：推銷產品，重在差異化，差別化，勝在差異化，差別化，切記。
A：發想…… B：目標…… C：願景……
A255: Selling products. Focus on differentiation. Differentiation. Winning in differentiation. Differentiation. Remember.
A: Think... B: Objective... C: Vision...

只要你願意，泥土也會變黃金。
As long as you want, the soil will become gold.

A256：服務的差異化，差別化，也是推銷必勝的要件，切記。
A：發想…… B：目標…… C：願景……
A256: Differentiation of services. Differentiation. It is also an essential requirement for sales promotion. Remember.
A: Think... B: Objective... C: Vision...

A257：凡是在失敗時，要試想著再給自己一個機會，再給自己一個努力，因為這是希望。
A：發想…… B：目標…… C：願景……
A257: Whenever it fails. Think about giving yourself another chance. Give yourself another effort. Because this is hope.
A: Think... B: Objective... C: Vision...

A258：是貧窮失敗，或是致富成功，要不要，決定權在你自己的手裡。
A：發想…… B：目標…… C：願景……
A258: It is poverty and failure. Or get rich successfully. Do you want. The decision is in your own hands.
A: Think... B: Objective... C: Vision...

A259：機會+行動+努力+持續=獲利。
A：發想…… B：目標…… C：願景……
A259: Opportunity + action + effort + continuity = profit.
A: Think... B: Objective... C: Vision...

A260：人生有條路可以走，但唯一走得通，走得順的，就必須絞盡腦汁，堅定信念，目標，決心，非致富成功不可。
A：發想…… B：目標…… C：願景……
A260: There is a way to go in life. But the only thing that works. Going

well. You must rack your brains. Firm conviction. aims. determination. You must succeed in getting rich.
A: Think... B: Objective... C: Vision...

A261：時時活用自己的腦袋，時時強化自己的技術和知識+技能，方可在市場上戰勝一切，在社會上發光發熱。
A：發想…… B：目標…… C：願景……
A261: Use your head all the time. Always strengthen your skills and knowledge + skills. Only then can you overcome everything in the market. In society, light is hot.
A: Think... B: Objective... C: Vision...

A262：改變心態，就是改變生命。
A：發想…… B：目標…… C：願景……
A262: Change your mindset. Is to change lives.
A: Think... B: Objective... C: Vision...

A263：市場是大的，客戶處處在你身邊，只要用心推廣，細心推廣，處處是機會，處處是有所獲。
A：發想…… B：目標…… C：願景……
A263: The market is big. Customers are all around you. Just push it hard. Carefully promote. Everywhere is an opportunity. Everywhere is something.
A: Think... B: Objective... C: Vision...

A264：路是人走出來的，市場是人開發出來的，隨時要堅信這個信念。
A：發想…… B：目標…… C：願景……
A264: The road came out of people. The market is developed by people.

A: Think... B: Objective... C: Vision...

A265：業務，公司，企業，定要把客戶擺在第一，人，事，物，擺在第一，這樣一來，事事就可以順利推動，事事就可水道渠成，事半功倍。
A：發想…… B：目標…… C：願景……
A265: Business. the company. enterprise. Must put the customer first. people. thing. Things. In the first place. Thus. Everything can be promoted smoothly. Everything can be done. Do more with less.
A: Think... B: Objective... C: Vision...

A266：給客戶最高的滿意度，永遠把客戶擺在第一位，這樣一來，業務就會蒸蒸日上，業務發展，就會大有可為。
A：發想…… B：目標…… C：願景……
A266: Give customers the highest satisfaction. Always put customers first. Thus. Business will flourish. business development. There will be great promise.
A: Think... B: Objective... C: Vision...

A267：強迫式的推銷，硬塞式的推銷，並不是好的推銷策略，而順應客戶的需求與期待，去解決客戶的需求與期待，才是最好的推銷策略。
A：發想…… B：目標…… C：願景……
A267: Forced sales. Hard plug sales. Not a good marketing strategy. And to meet customer needs and expectations. To solve customer needs and expectations. Is the best marketing strategy.
A: Think... B: Objective... C: Vision...

A268：專業技術+專業知識+EQ=成功

A268: Professional technology + professional knowledge + EQ = success

A269：井易免店面飲水機，加盟事業，是一個很好的加盟事業，也可以讓你很輕鬆的學會這個事業的技能，經營，發展，有前瞻的事業，而這個事業的市場，商機，到處都是市場，到處都是商機，讓你很輕易就可以賺進你渴望想要的財富。

A：發想…… B：目標…… C：願景……

A269: Jingyi free store water dispenser. Join the cause. Is a very good franchise business. It also allows you to learn the skills of this career very loosely. Operation. development of. Prospective career. And the market for this business. Business. There are markets everywhere. Business opportunities are everywhere. Make it easy for you to earn the wealth you desire.

A: Think... B: Objective... C: Vision...

A270：做生意，跑業務的成功，95%是情緒+心態，成功的關鍵在情緒+心態+熊度。

A：發想…… B：目標…… C：願景……

A270: Doing business. Run business success. 95% is emotional + mentality. The key to success is emotion + mentality + bear degree.

A: Think... B: Objective... C: Vision...

A271：思考的力量，遠大於整天在埋頭苦幹，給自己有更多的思考空問，思考就力量。

A：發想…… B：目標…… C：願景……

A271: The power of thinking. Far more than working hard all day long. Give yourself more thinking space. Thinking is power.

A: Think... B: Objective... C: Vision...

情緒+態度=成功

Emotion + attitude = success

A272：對重點客戶，偶而或是定期贈一點小禮物，這樣一來，有助於人際關係和人際連結關係，有助於銷售量大增，或人際關係的增長，而進而增進與客戶的友好和信賴關係。

A：發想…… B：目標…… C：願景……

A272: For key customers. Occasionally or regularly give small gifts. Thus. Help interpersonal relationships and interpersonal connections. Help increase sales. Or the growth of interpersonal relationships. In turn, the friendly and trusting relationship with customers is improved.

A: Think... B: Objective... C: Vision...

A273：戲法人人會變，各有巧秒不同，用思維，改變思維，未來就不是夢。

A：發想…… B：目標…… C：願景……

A273: Everyone will change the trick. Every second is different. Use thinking. Change thinking. The future is not a dream.

A: Think... B: Objective... C: Vision...

A274：本書所有的敘述，都在培養出優質的菁英人士.來服務消費大眾，來服務社會更和諧，更進步，更有競爭力。

A：發想…… B：目標…… C：願景……

A274: All the narratives in this book. All are cultivating high-quality elites to serve consumers. To serve the society more harmoniously. More progress. is more competitive.

A: Think... B: Objective... C: Vision...

A275：不斷前進，不斷進步，不斷改造，不斷創新，迅速進步。

A：發想…… B：目標…… C：願景……

A275: Keep moving forward. Keep going. Constant transformation. innovation. Enter quickly.

A: Think... B: Objective... C: Vision...

A276：客戶在那裡，我們的服務就在那裡。
A：發想…… B：目標…… C：願景……
A276: The customer is there. Our service is there.
A: Think... B: Objective... C: Vision...

A277：會賣沒有什麼了不起，談個好價錢，才了不起。
A：發想…… B：目標…… C：願景……
A277: The power of thinking. Far more than working hard all day long. Give yourself more thinking space. Thinking is power.
A: Think... B: Objective... C: Vision...

A：278：千萬追夢計劃，自己的命運由你自己主導。
A：發想…… B：目標…… C：願景……
A: 278: Ten million dream chase plan. Your own destiny is dominated by yourself.
A: Think... B: Objective... C: Vision...

A279：要達成目標，務必堅定自己的信念，肯定自己，相信自己，要有超強的自信心，方可走到目標，達成目標。
A：發想…… B：目標…… C：願景……
A279: To achieve the goal. Make sure your faith. Affirm yourself. trust yourself. Must have super self-confidence. Only to reach the goal. goal achieved.
A: Think... B: Objective... C: Vision...

A280：隨著時空的變遷，業務員隨時隨地，腦筋要大轉動，迎合時空的變遷，跟緊時空的腳步，創造最大的商機。

A：發想…… B：目標…… C：願景……
A280: With the change of time and space. Salesman anytime, anywhere. The brain is going to turn a lot. Cater to changes in time and space. Keep pace with time and space. Create the biggest business opportunities.
A: Think... B: Objective... C: Vision...

A281：戰鬥的心，不會因為任何的因素而停頓，必須大力奮發，勇往直前，持續戰鬥，永不停止，永不放棄業務員應具備的心態。
A：發想…… B：目標…… C：願景……
A281: The heart of battle. Will not stop for any reason. Must work hard. Go forward bravely. Continue to fight. Never stop. Never give up the mentality that salesmen should possess.
A: Think... B: Objective... C: Vision...

A282：勇氣，毅力，戰鬥的心，是支撐你成功的要件。
A：發想…… B：目標…… C：願景……
A282: Brave. perseverance. Fighting heart. It is the key to your success.
A: Think... B: Objective... C: Vision...

A283：每天要用 100%的心態，去面對工作，面對事業，讓每天發光發亮發熱。
A：發想…… B：目標…… C：願景……
A283: Use 100% mentality every day. To face work. Face career. Let every day shine and heat.
A: Think... B: Objective... C: Vision...

A284：接待客戶的熱情要不斷的提升，服務客戶的滿意度要不斷的持續的提升，接待，服務，滿意度要做到最好，最完美，最極致。
A：發想…… B：目標…… C：願景……
A284: The enthusiasm of receiving customers should be continuously improved. The satisfaction of service customers should be continuously improved. Reception. service. Satisfaction must be the best. perfect. The most extreme.
A: Think... B: Objective... C: Vision...

A285：材料，技術，資料，資訊，銷售要齊全，方能事半功倍。
A：發想…… B：目標…… C：願景……
A285: Materials. technology. data. Information. Sales must be completed. It can do more with less.
A: Think... B: Objective... C: Vision...

A286：失敗的人，在機會中找藉口或找困難，成功的人，在困難中找機會，或抓住機會。
A：發想…… B：目標…… C：願景……
A286: The failed person. Find excuses or difficulties in opportunities. successful people. Find opportunities in difficulties. Or seize the opportunity.
A: Think... B: Objective... C: Vision...

A287：待人處事，或人際關係，要真誠，要熱忱，要熱情，更要與人合群相處，正向的態度和正向的人生價值觀。
A：發想…… B：目標…… C：願景……
A287: Treat others. Or interpersonal relationships. Be sincere. Be enthusiastic. Be enthusiastic. It is more important to get along with people. Positive attitude and positive life values.
A: Think... B: Objective... C: Vision...

人脈是致富的跳板，人脈不一定是錢脈，而是要有正確的人脈，才正是錢脈。
The network is a springboard for getting rich. The network is not necessarily the money. It is the right network. It is the money.

A288：五品，一認錯二柔和三樂忍四溝通五放下，機會+行動+努力+持續=獲利。黃金成功定律：1.目標 2.計劃 3.設定期限 4.立即行動 5.早晚大聲朗誦 6.持之以恆。
A：發想…… B：目標…… C：願景……
A288: Fifth grade. One admits mistakes, two is soft, three is happy, four communicates, and five puts down. Opportunity + action + effort + sustainability = profit. The Golden Law of Success: 1. Goal 2. Plan 3. Set deadline 4. Act now 5. Speak aloud sooner or later 6. Persevere.
A: Think... B: Objective... C: Vision...

A289：挫折和失敗，是你未來成功的墊腳石，千萬要記住，不是你未來成功的絆腳石，這是自然的真理。
A：發想…… B：目標…… C：願景……
A289: Frustration and failure. Stepping stone for your future success. Do remember. Not a stumbling block to your future success. This is natural truth.
A: Think... B: Objective... C: Vision...

A290：當客戶遇到任何問題或業務員，你要儘可能的，即時的，協助客戶解決問題或困難，這樣一來，不但可以守住客戶，更可以增進與客戶彼此的友誼和信賴關係。
A：發想…… B：目標…… C：願景……
A290: When the customer encounters any problems or salesman you. As much as possible. real-time. Assist customers to solve problems or difficulties. Thus. Not only can we keep our customers. It can also enhance the friendship and trust relationship with customers.
A: Think... B: Objective... C: Vision...

A291：正面的思維，正向的態度，正能量的價值觀，可強化你的推銷意識，強化你的銷售信念。

A：發想…… B：目標…… C：願景……

A291: Positive thinking. Positive attitude. Positive energy values. Can strengthen your sales awareness. Strengthen your sales beliefs.

A: Think... B: Objective... C: Vision...

A292：一步一腳印，走自己的路，走自己的生活.做最興趣的事，做自己最擅長的工作、把它做到最好，做到最極致，這樣你的事業就比較會成功，比較會快樂，創造最美好的人生。

A：發想…… B：目標…… C：願景……

A292: One step at a time. walk my own path. Take your own life. Do the things you are most interested in. Do what you do best, and do it best. Do the most extreme. Then your career will be more successful. Would be happier. Create the best life.

A: Think... B: Objective... C: Vision...

A293：釋放無限潛能，創造最大商機，創造最大財富，創造最美好生活。

A：發想…… B：目標…… C：願景……

A293: Unleash unlimited potential. Create the biggest business opportunities. Create the greatest wealth. Create the best life.

A: Think... B: Objective... C: Vision...

A294：相信自己，肯定自己，只要用心，只要奮鬥持續前進，相信絕對能能創造最大成就和最大財富，創造最美好的生活。

A：發想…… B：目標…… C：願景……

A294: Believe in yourself. Affirm yourself. Just work hard. As long as the struggle continues. I believe it can definitely create the greatest achievements and the greatest wealth. Create the best life.

樂觀的人，永遠看到問題背後的機會。

Optimistic person. Always see the opportunity behind the problem.

A: Think... B: Objective... C: Vision...

A295：常常鼓勵人家，常常讚美人家，常常肯定人家，常常重視人家，讓人家感覺你對人家很重視，很重要，很尊重，因為都希望得到別人的肯定和尊重，這是待人處世最基本的常識和概念，使人與人的人際關係變得圓融更好。
A：發想…… B：目標…… C：願景……
A295: People are often encouraged. Praise people often. Often people are sure. Often value people. You feel that you value others very much. Very important. Very respectful. Because they all want to be recognized and respected by others. This is the most basic common sense and concept of dealing with others. Make people's interpersonal relationships become better and better.
A: Think... B: Objective... C: Vision...

A296：我們的服務，遠比你想像的還要多。
A：發想…… B：目標…… C：願景……
A296: Our service. Far more than you think.
Athought... B goal... C Vision...

A297：服務是銷售的開始，銷售是服務的延續。
A：發想…… B：目標…… C：願景……
A297: Service is the beginning of sales. Sales are a continuation of service.
A: Think... B: Objective... C: Vision...

A298：只要你肯相信自己，只要你想要，你就會擁有一切，只要你想要，凡是工作，事業，財富，就會像雪片般的一片一片飄過來，你想擋，都擋不住，這是自然不變的定律。

A：發想…… B：目標…… C：願景……

A298: As long as you believe in yourself. As long as you want. You will have everything. As long as you want. All work. cause. Wealth animal. It will float like a piece of snow. You want to block it. Can't stop it. This is a natural and unchanging law.

A: Think... B: Objective... C: Vision...

A299：多一個朋友，多一道路，少一個朋友，多一道牆，廣結善緣，結交順友，結交社群，這樣有助於業務。

A：發想…… B：目標…… C：願景……

A299: One more friend. One more way. One less friend. One more wall. Wide friendship. Make good friends. Make a community. This helps the business.

A: Think... B: Objective... C: Vision...

A300：技術是保障，知識是力量，隨時強化技術，隨時增進知識，使人生更豐富，更多采多姿，更美好絢麗。

A：發想…… B：目標…… C：願景……

A300: Technology is the guarantee. Knowledge is power. Strengthen technology at any time. Improve knowledge at any time. Make life richer. More colorful. More beautiful and gorgeous.

A: Think... B: Objective... C: Vision...

A301：正向的態度，正向的人際人關係，負面的態度，負面的人際關係，多讚賞，多肯定，多讚美，多美言，多傾聽，好相處，好合群，這都是正向的態度和正向的人際，愛抱怨，難相處，難合群，難溝通，不合群，不協助，不傾聽，這都是負面的熊和負面的人際關係，正面態度和負面態度，正面人際關係和負面際關係，這對你未來事業和人生影響深遠。

A：發想…… B：目標…… C：願景……

A301: Positive attitude. Positive interpersonal relationships. Negative attitude. Negative relationships. Much appreciated. How sure. Much praise. What a beautiful word. Listen more. Get along well. Good group. This is a positive attitude and positive interpersonal relationship. Love to complain. Disagreeable. Difficult to group. Difficult to communicate. Not social. Does not assist. Don't listen. These are all negative bears and negative relationships. Positive attitude and negative attitude. Positive relationships and negative relationships. This will have a profound impact on your future career and life.
A: Think... B: Objective... C: Vision...

A302：一個對的念頭堅持下去，一個好的念頭，就能成就大事。
A：發想…… B：目標…… C：願景……
A302: Hold on to a right idea. Agood idea. Can accomplish great things.
A: Think... B: Objective... C: Vision...

A303：對自己要信心拾足，要全力衝刺，全力以赴，對事的成功要信心拾足，對目標要信心拾足，對未來要信心拾足，要全力衝刺，全力以赴，方可創造美好的未來。
A：發想…… B：目標…… C：願景……
A303: Be confident + sufficient in yourself. Do your best to sprint. Go all out. Confidence in the success of things + enough. Be confident + full of goals. Be confident + full in the future. Do your best to sprint. Go all out. Only to create a better future.
A: Think... B: Objective... C: Vision...

A304：錯過的機會，要找回來很難，所以掌握即時的機會，掌握當下的機會很重要，也很必要，這是生意事業成功的關鍵所在。

A：發想…… B：目標…… C：願景……

A304: Missed opportunities. It's hard to get it back. So grasp the immediate opportunity. It is important to grasp the opportunities of the moment. It is also necessary. This is the key to business success.

A: Think... B: Objective... C: Vision...

A305：勇者跟懦者的差別，勇者面對問題，解決問題，懦者逃避問題，退縮問題。

A：發想…… B：目標…… C：願景……

A305: The difference between the brave and the coward. The brave face the problem. Solve the problem. Coward. Escape the problem. Retreat.

A: Think... B: Objective... C: Vision...

A306：做人在前，做事在後，做人在前，做生意在後。

A：發想…… B：目標…… C：願景……

A306: Be the first to be a person. Do the last to be a person.

A: Think... B: Objective... C: Vision...

A307：要好好珍惜每一個客戶，要好好珍惜每一筆生意，每一個客戶，每一筆生意，都是事業發展的保障，事業起飛的翅膀。

A：發想…… B：目標…… C：願景……

A307: Cherish every customer. Cherish every business. Every customer. Every business. It is the guarantee of career development. The wings of career take-off.

A: Think... B: Objective... C: Vision...

A308：每一個客戶，都要做的很週到，很完整，每一個客戶都要服務的很完整，每一筆生意都要很珍惜，很寶貴，這會倍增業務績效，切記。
A：發想…… B：目標…… C：願景……
A308: Every customer has to be very thoughtful. Very complete. Every customer must serve very well. Every business must be cherished. Very precious. This will multiply the business efficiency. Remember.
A: Think... B: Objective... C: Vision...

309：知道沒有力量，相信才有力量，知道只是啟發，執行才有力量。
A：發想…… B：目標…… C：願景……
A309: Know that there is no power. Believe in power. Knowing is only enlightening. Execution has power.
A: Think... B: Objective... C: Vision...

A310：經驗才能解決問題，理論只能解答問題，經驗可以抉擇問題，理論只能提供參考。
A：發想…… B：目標…… C：願景……
A310: Experience can solve problems. Theory can only answer problems. Experience can choose problems. Theory can only provide reference.
A: Think... B: Objective... C: Vision...

A311：一個笑容，帶來一大筆生意，一個臭臉，失去一大筆生意，一個關心，帶來一大筆生意，一個冷漠，失去一片天空，這是天經地義，不變的真理。
A：發想…… B：目標…… C：願景……
A311: A smile. Bring a lot of business. Abad face loses a lot of business. One care. One brings a lot of business. An indifferent loses a sky.
A: Think... B: Objective... C: Vision...

華人獨特的人際關係：
A.認親戚 B.拉關係 C.攀交情 D.做人情 E.鑽營 F.送禮。
C35: The unique interpersonal relationship of the Chinese:
A. Recognize relatives B. Pull relationships C. Cultivate friendship
D. Do human relations E. Drill camp F. Give gifts.

A312：多認識一個人，多認識一個客戶，多熟悉一個人，事，時，地，物，這是做生意，做事業的緣分，機會，機遇，要好好珍惜，也是做生意，做事業，最好的培養土。

A：發想…… B：目標…… C：願景……

A312: Know one more person. One more customer. One more person. Things. Time. Place. Things. This is business. Business. The best cultivation soil.

A: Think... B: Objective... C: Vision...

A313：時間用在那裡，成就就在那裡，投資在那裡，收穫就在那裡。

A：發想…… B：目標…… C：願景……

A313: Time is there. Achievements are there. Investments are there. Harvests are there.

A: Think... B: Objective... C: Vision...

A314：愛心是人類最大的力量，情感是人類最大的影響力，有愛心就有人脈，有情感就有豐富的人生。

A：發想…… B：目標…… C：願景……

A314: Love is the greatest power of human beings. Emotion is the greatest influence of human beings. There is love. There are contacts.

A: Think... B: Objective... C: Vision...

A315：命運靠自己扭轉，扭轉才能找到希望，找到未來。

A：發想…… B：目標…… C：願景……

A315: Fate is reversed by oneself. Reversal can find hope. Find the future.

A: Think... B: Objective... C: Vision...

先認識人，再交流，再交心，這是交朋友最佳上策。

Meet people first. Then communicate. Then make friends. This is the best way to make friends.

A316：掌握即時的機會，抓住瞬間的機會，是成功致富的關鍵所在，掌握即時的機會，是創造財富的保證。

A：發想…… B：目標…… C：願景……

A316: Master the immediate opportunity. Seize the opportunity of Shunjian. It is the key to success and get rich. Master the instant opportunity. It is the guarantee of wealth creation.

A: Think... B: Objective... C: Vision...

A317：年青人用錯方法，難成功，要選對創業平台，掌握創業平台，這樣不成功也難，切記。

A：發想…… B：目標…… C：願景……

A317: Young people use the wrong method. It is difficult to succeed. It is necessary to choose the right entrepreneurial platform. Master the entrepreneurial platform. It is difficult to be unsuccessful.

A: Think... B: Objective... C: Vision...

A318：千百次奮鬥，不如一次選擇錯誤，經驗是幫助大家飛向成功的翅膀，切記。

A：發想…… B：目標…… C：願景……

A318: Thousands of struggles. It is better to choose wrong one time. Experience is to help everyone fly to the wings of success. Remember.

A: Think... B: Objective... C: Vision...

A319：危機危險的背後，就是機會，抓住機會，締造不可能的「可能」這兩個字。

A：發想…… B：目標…… C：願景……

A319: Behind the danger of crisis. Opportunity. Seize the opportunity. Create the impossible "possible" word.

A: Think... B: Objective... C: Vision...

A320：啟迪人心，激發潛能，對未來懷抱著熱忱熱情，未來成功指日可待。

A：發想…… B：目標…… C：願景……

A320: Enlighten people's hearts. Inspire potential. Enthusiasm for the future. Future success is just around the corner.

A: Think... B: Objective... C: Vision...

A321：啟發全方位人生，勇敢改變價值觀，勇敢改變人生觀。

A：發想…… B：目標…… C：願景……

A321: Inspire all-round life. Brave to change values. Brave to change the outlook on life.

A: Think... B: Objective... C: Vision...

A322：開拓人生視野，探討事業機會，透過遠見和國際視野，改變我們的生活。

A：發想…… B：目標…… C：願景……

A322: Open up the horizons of life. Explore career opportunities. Change our lives through vision and international vision.

A: Think... B: Objective... C: Vision...

A323：一個人做人做事，要隨緣，好相處，多付出，就多人緣，得人脉，得人群。

A：發想…… B：目標…… C：願景……

A323: One person. Being a person. Doing things with others. Get along well. Get along well. Pay more. Just lose more people. Get connections. Get crowds.

A: Think... B: Objective... C: Vision...

一個人的成功，15%取決於專業技術，85%取決於人際溝通能力。

One′s success. 15% depends on professional skills. 85% depends on interpersonal communication skills.

A324：遇有機會，就要推銷，處處要推銷.隨時要推銷，這樣生意.才能蒸蒸日上，日日興隆。

A：發想…… B：目標…… C：願景……

324: If there is a chance, it must be promoted. It must be promoted everywhere. It must be promoted at any time. In this way, the business can flourish.

A: Think... B: Objective... C: Vision...

A325：做人做事，做生意，要努力發揮創造力，要努力經營，良好的資源，豐富的人脈，賺錢的機會，要好好的握住，好好珍惜，創造多贏的環境。

A：發想…… B：目標…… C：願景……

A325: To be a person, to do business, to do business, to work hard to be creative, to work hard, to have good resources, to have rich connections, to make money, to hold on, to cherish, to create a win-win environment.

A: Think... B: Objective... C: Vision...

A326：一個人的成功，必須先從觀念和認知改變，再改變行動和習慣，做整個結構性的改變，改變自己，全力以赴。

A：發想…… B：目標…… C：願景……

A326: A person's success must first change from concept and cognition. Then change the actions and habits. Make the whole structural change. Change yourself. Go all out.

A: Think... B: Objective... C: Vision...

A327：一個人的成功，不在於能力強不強，而在於信念好不好，不在於修行強不強，而在於觀念好不好，目標設定，全力以赴。

A：發想…… B：目標…… C：願景……

A327: A person's success. It is not in the ability to be strong. It is in the belief or not. It is not in the practice or not. It is in the concept. The goal is set. Go all out.
A: Think... B: Objective... C: Vision...

A328：思路決定你的出路，觀念決定你的貧富，眼光決定你的未來。
A：發想…… B：目標…… C：願景……
A328: Thinking determines your way out. Concept determines your rich and poor. Vision determines your future.
A: Think... B: Objective... C: Vision...

A329：做人做事，要有大格局，格局有多大，事業就有多大，時時培養與人合作的能力，方可達成全方位的成功，全方位的成功人生。
A：發想…… B：目標…… C：願景……
A329: To be a person. To have a big pattern. How big the pattern is. How big a career is. Cultivate the ability to cooperate with people all the time.
A: Think... B: Objective... C: Vision...

A330：有行動力，就會有無限可能發生，人在生活中，隨時隨地要去強化自己的生活能力與能量。
A：發想…… B：目標…… C：願景……
A330: There is action. There will be infinite possibilities. People in life. Anytime, anywhere. To strengthen their ability and energy in life.
A: Think... B: Objective... C: Vision...

A331：做事情，做事業，必須具備：1.時間目標 2.事務目標 3.願望目標 4.目標導向，必須一一的去完成，達成，方可創造事業高峯。

A：發想…… B：目標…… C：願景……

A331: Do things. Do business. Must have: 1. Time goals 2. Business goals 3. Wish goals 4. Goal-oriented. Must be completed one by one. Achieved in order to create a business peak.

A: Think... B: Objective... C: Vision...

A332：人生有許多苦，轉念就是樂，最努力的人，最容易成功，最最最努力的人，最最最能成功。

A：發想…… B：目標…… C：願景……

A332: There are a lot of hardships in life. Change is joy. The hardest person. The easiest to succeed. The most hardworking. The most successful.

A: Think... B: Objective... C: Vision...

A333：不斷的創新，不斷的改變，不斷的努力，敢拼，敢做，一直不放棄，就能成功。

A：發想…… B：目標…… C：願景……

A333: Constant innovation. Constant change. Constant effort. Dare to fight. Dare to do. Never give up. Can succeed.

A: Think... B: Objective... C: Vision...

334：不敗的勇者，就是鬥志一直在。

A：發想…… B：目標…… C：願景……

A334: The undefeated brave. It is the fighting spirit that has always been there.

A: Think... B: Objective... C: Vision...

A335：人不理財，財就不理人，人要理財，財才會理人。

A：發想…… B：目標…… C：願景……

A335: People don't manage money. People don't care about money. People want to manage money.
A: Think... B: Objective... C: Vision...

A336：失敗是檢討的動能，失敗也是前進的動能，失敗也是邁向再成功的動能。
A：發想…… B：目標…… C：願景……
A336: Failure is the kinetic energy of review. Failure is also the kinetic energy of progress. Failure is also the kinetic energy of moving towards success.
A: Think... B: Objective... C: Vision...

A337：做生意，做事業，若有姑且一試的心態，那是不會成功的，必須有全力以赴的態度，才能會有成功的。
A：發想…… B：目標…… C：願景……
A337: Doing business. Doing a business. If you have the mentality to give it a try, it will not succeed. You must have an attitude of doing your best.
A: Think... B: Objective... C: Vision...

A338：走舊路，就走不到新地方，走舊思維，就走不到新的創意。
A：發想…… B：目標…… C：願景……
A338: Take the old road. You can't go to new places. Take old ideas. You can't go to new ideas.
A: Think... B: Objective... C: Vision...

A339：做事要用心，才能成功，做每一件事都要用心，才能有所成長，才能一天比一天更好，也才能成功。
A：發想…… B：目標…… C：願景……

熱愛你的工作，熱愛你的朋友，熱愛你的生活，這是人生成功要素。
Love your work. Love your friends. Love your life. This is the key to success in life.

A339: It takes effort to do things. To be successful. To do everything is to be attentive. To grow, to be better every day. To be successful.
A: Think... B: Objective... C: Vision...

A340：人潮，錢潮，是人脈的存摺，愈多就愈有發展，也就是發展的基石。
A：發想…… B：目標…… C：願景……
A340: Crowd of people. Tide of money. It is the passbook of connections. The more it develops, the more it is the cornerstone of development.
A: Think... B: Objective... C: Vision...

A341：對的事情，對的事業，要敢想，要敢做，要敢衝，要敢做換位思考，要敢做逆向思考，要敢做事業大夢。
A：發想…… B：目標…… C：願景……
A341: The right thing. The right business. Dare to think. Dare to do. Dare to rush. Dare to do empathy. Dare to do reverse thinking. Dare to dream big.
A: Think... B: Objective... C: Vision...

A342：創新是進步的原動力，把夢想化為行動，用行動來實現夢想。
A：發想…… B：目標…… C：願景……
A342: Innovation is the driving force behind progress. Turn dreams into actions. Use actions to realize dreams.
A: Think... B: Objective... C: Vision...

A343：好的行銷，好的產品，終究遲早會被發現，其策略，要不斷的宣傳，不斷的做普羅大眾的教育訓練，比如，一年做普羅大眾教育訓練三百場等等。

A：發想…… B：目標…… C：願景……

A343: Good marketing. Good products. Sooner or later, it will be discovered. Its strategy. Constant publicity. Constant education and training of the general public. For example, 300 training courses of general public a year, etc.

A: Think... B: Objective... C: Vision...

A344：有行動力，不管你年齡大小，生活就會有希望，生命就會價值。

A：發想…… B：目標…… C：願景……

A344: There is action. No matter what your age is, life will have hope. Life will be worth.

A: Think... B: Objective... C: Vision...

A345：做生意，處理人際面，人際關係，不要因為小事而壞了大事，不要因為小局而壞了大局，不要因為小錢而壞了大錢，而壞了大事，千萬要記住。

A：發想…… B：目標…… C：願景……

A345: Doing business. Dealing with interpersonal relationships. Interpersonal relationships. Do n't choke on big things because of small things. Do n't choke on big things because of small things.

A: Think... B: Objective... C: Vision...

A346：客戶對你的批評，不屑，不禮貌，都要虛心接受，千萬不可動怒，生氣，否則你會失去一切，千萬要記住。

A：發想…… B：目標…… C：願景……

A346: Customers criticize you. Disdain. Disrespectful. Accept with humility. Do n't be angry. Be angry. Otherwise you will lose everything. Do n't forget.

A: Think... B: Objective... C: Vision...

好的溝通能力+好的人際關係，是事業成功的關鍵所在。

Good communication skills + good interpersonal relationships. It is the key to successful career.

A347：我最好，我相信，我可以，我相信，我能成功，我相信，要時時培養強烈的自信心。
A：發想…… B：目標…… C：願景……
A347: I'm the best. I believe. I can. I believe. I can succeed. I believe. Always cultivate strong self-confidence.
A: Think... B: Objective... C: Vision...

A348：勇敢的去做，才能發現新的機會，新的可能.新的希望。
A：發想…… B：目標…… C：願景……
A348: Do it bravely. Only in this way can you discover new opportunities, new possibilities, new hopes.
A: Think... B: Objective... C: Vision...

A349：全球競爭力，決定在學習力，而學習力決定在行動力。
A：發想…… B：目標…… C：願景……
A349: Global competitiveness. Decide on learning ability. And learning ability decides on action ability.
A. Thinking... B. Goal... C. Vision...

A350：積極面對所有的挑戰，立即行動終生堅持。
A：發想…… B：目標…… C：願景……
A350: Actively face all challenges. Act immediately and persist for life.
A: Think... B: Objective... C: Vision...

A351：隨時啟動心靈的力量.瞬間改變你全方位的人生。
A：發想…… B：目標…… C：願景……
A351: Start the power of the mind at any time. Instantly change your all-round life.
A: Think... B: Objective... C: Vision...

怎麼樣去發掘和開發自己潛在的能量，相信自己，肯定自己，自己是可以的，這是潛能開發最大的推動力。
How to discover and develop your potential energy. Believe in yourself, affirm yourself, and be yourself. This is the biggest driving force for potential development.

A352：怎麼樣才能賺大錢？1.技術要好 2.知識要好 3.要有智慧 4 銷售第一 5.量要大 6.產品品質要好 7.服務要第一 8.市場要大要第一 9.人才素質要第一 10.人脈要廣，要廣結善緣，這樣就能賺大錢。

A：發想…… B：目標…… C：願景……

A352: How to make big money? 1. Technology is better 2. Knowledge is better 3. Wisdom is required 4. Sales are first 5. Quantity is large 6. Product quality is better 7. Service is first 8. Market is big and first. 9. The quality of the talents should be the first 10. The network should be wide. The good affiliation should be made widely.

A: Think... B: Objective... C: Vision...

A353：努力+機會.就能看到發展，努力+機會，就能看到用心，堅持成功，自然看得見。

A：發想…… B：目標…… C：願景……

A353: Efforts + Opportunities. You can see development. Efforts + Opportunities. You can see intentions. Persist in success. Naturally visible.

A: Think... B: Objective... C: Vision...

A354：把簡單的事，做到最好，自然就能脫穎而出，自然就能邁向成功之路，把小事當大事來做，自然就可以做到盡善盡美，勝利成功。

A：發想…… B：目標…… C：願景……

A354: Do the simple things. Do the best. Naturally you can get out of the way. Naturally you can make your way to success. Do small things as big things. Naturally you can achieve perfection. Victory and success.

A: Think... B: Objective... C: Vision...

常常給予真誠的讚美和感謝，這是友誼和人際關係昇華的必備元素。

Always give sincere praise and gratitude. This is an essential element for the sublimation of friendship and interpersonal relationships.

A355：知識=財富，技能=財富，經驗=財富，有知識，有技能，有經驗，財富自然就來。

A：發想…… B：目標…… C：願景……

A355: Knowledge = wealth. Skills = wealth. Experience = wealth. Knowledge. Skills. Experience. Wealth comes naturally.

A: Think... B: Objective... C: Vision...

A356：要改變自己的命運，先改變自己的心態和態度，才能改變自己的命運。

A：發想…… B：目標…… C：願景……

A356: To change one's destiny. First change one's mentality and attitude. Then one can change one's own destiny.

A: Think... B: Objective... C: Vision...

A357：每一個小客戶，都要把他當做大客戶來對待，來服務，這樣客戶才會源源不斷。

A：發想…… B：目標…… C：願景……

A357: Every small customer must treat him as a big customer. Come to serve. Only in this way will customers continue to flow.

A: Think... B: Objective... C: Vision...

A358：推銷術或行銷術，就是等於不斷持續的推銷，大步，大力，不停的前進推銷，總是會有機會碰到有緣人，碰到大客戶，這樣你的人生就從此改變了，從此邁向成功之路。

A：發想…… B：目標…… C：願景……

A358: Salesmanship or marketing. It is equal to continuous salesmanship. Big step. Vigorously. Non-stop advancing salesmanship. There will always be a chance to meet someone with love. Encounter with a big customer. Just like this, your life will change from then on. Then. From then on to the road to success.

常常引發他人心中的渴望，傾聽他人心中的需求，而給予充分的協助和幫助。

Frequently arouse the desires of others. Listen to the needs of others and give them full assistance and help.

A: Think... B: Objective... C: Vision...

A359：品質，技能，研發，態度，是個人或企業決勝的關鍵，也是個人或企業成功或失敗的命脈。

A：發想…… B：目標…… C：願景……

A359: Quality. Skills. R & D. Attitude is the key to success of individual or enterprise. It is also the lifeblood of success or failure of individual or enterprise.

A: Think... B: Objective... C: Vision...

360：有底子，就會有路子，只要基礎打的好，生意，事業，自然就會有預想不到的好發展.好榮景。

A：發想…… B：目標…… C：願景……

A360: There is a foundation. There will be a way. As long as the foundation is well established. Business. Career. Naturally, there will be unexpectedly good development. Good glory.

A: Think... B: Objective... C: Vision...

A361：要把小客戶當大客戶來做，來服務，來尊重，把小客戶當大客戶來照顧，使其尊榮，使其榮耀。

A：發想…… B：目標…… C：願景……

361: Small customers should be treated as large customers, service should be provided, respect should be taken, and small customers should be taken care of as large customers.

A: Think... B: Objective... C: Vision...

A362：你自己的未來，掌握在你自己的手中，相信自己，肯定自己，大步勇敢前進，成功就在現在，成功就在眼前，抓住現在，抓住眼前，你的未來就不是夢。

A：發想…… B：目標…… C：願景……

A362: Your own future. Hold it in your own hands. Believe in yourself. Affirm yourself. Bravely advance. Success is now. Success is in sight. Seize the present. Hold your eyes. Your future is not a dream.

A: Think... B: Objective... C: Vision...

A363：人生是福，是禍，是成功，是失敗，是快樂，是不快樂，取決於「轉念」這兩個字，很重要，是你人生的轉折點。

A：發想…… B：目標…… C：願景……

A363: Life is a blessing. It is a disaster. It is a success. It is a failure. It is happiness. It is unhappiness. It depends on the word. Very important. It is a turning point in your life.

A: Think... B: Objective... C: Vision...

A364：成功的基本十個概念：1.細心+熱情=成功 2.夢想+實踐=成功 3.勇敢+邁進=成功 4.發想+永恆的奮鬥=成功 5.思考+決策=成功 6.思考+分析=成功 7.想法+行動力=成功 8.想法+執行力=成功 9.心態+人際關係=成功 10.態度+人際關係=成功

A：發想…… B：目標…… C：願景……

A364: The ten basic concepts of success: 1. Carefulness + enthusiasm = success 2. Dream + practice = success 3. Bravery + stride = success 4. Ideas + eternal struggle = success 5. Thinking + decision making = success 6. Thinking + Analysis = Success 7. Idea + Mobility = Success 8. Idea + Execution = Success 9. Mindset + Interpersonal Relationship = Success 10. Attitude + Interpersonal Relationship = Success

A: Think... B: Objective... C: Vision...

A365：凡是一切事物，機會是留給有準備的人，只要堅持再堅持，積極面對所有的挑戰，立即行動終生堅持，只要不妄自菲薄，不放棄，努力奮鬥向前，成功就愈來離你愈近了，這是不

小客戶，要當大客戶來服務，來經營，這是業務工作，成功的基本概念。

Small customers. To serve as big customers. To operate. This is the basic concept of business work. Success.

變的自然定律，也是不變的自然法則，願大家一切幸福美滿，願大家一切成功發財。

A：發想…… B：目標…… C：願景……

A365: Everything is everything. Opportunity is reserved for those who are prepared. Just persevere, persevere, face all challenges positively, act immediately and persevere forever, as long as you do n't arbitrarily, do n't give up, work hard and move forward, success will be more The closer you are. This is the unconstrained law of nature. It is also the unchanging law of nature. May everyone be happy and happy. May everyone be successful in making a fortune.

A: Think... B: Objective... C: Vision...

B.圖樣範例 Sample Picture Templat

想想看……想像一下……寫下來……

Think about it... imagine... write it down...

圖 1. 掌握全球資訊就是力量……

Figure 1. Mastering global information is power...

愛因斯坦說，一個人的成功常取決於轉折點上。

Einstein said that one′s success often depends on the turning point.

圖 2. 創意、創新，就是力量……
A：發想…… B：目標…… C：願景……
Figure 2. Creativity, innovation is power...
A: Think... B: Objective... C: Vision...

成功的銷售或服務，就是瞭解並做到超越客戶的期待和期望。
Successful sales or service. It is to understand and exceed customer expectations and expectations.

圖 3. 服務就是力量……
A：發想…… B：目標…… C：願景……
Figure 3. Service is power...
A: Think... B: Objective... C: Vision...

隨時檢視自己的工作效率，想辦法倍增工作效率。
Check your work efficiency at any time. Find ways to increase work efficiency.

圖 4. 冰山理論，發揮潛能，就是力量……
A：發想…… B：目標…… C：願景……
Figure 4. Iceberg theory. Realizing your potential. It's power...
A: Think... B: Objective... C: Vision...

銷售之所以成功，就是能掌握市場的需求，大量的行銷概念。
The reason why sales are successful is to be able to grasp the needs of the market. A large number of marketing concepts.

圖 5. 青蛙效應，要有危機意識……
A：發想…… B：目標…… C：願景……
Figure 5. The frog effect. Have a sense of crisis.
A: Think... B: Objective... C: Vision...

隨時掌握最新的資訊，隨時檢討，分析，做決策，納入公司決策一環，並確實實行之。

Keep up to date with the latest information at any time. Review, analyze, make decisions at any time, and incorporate it into the company's decision-making process.

圖 6. 危機意識，化危機為轉機，轉念就是力量……
A：發想…… B：目標…… C：願景……
Figure 6. Crisis consciousness. Turning crisis into turning point. Turning thoughts is power...
A: Think... B: Objective... C: Vision...

隨時分析自己的優點和缺點，隨時擴大優點，縮減缺點，使自己變得更優質，更卓越，更精進。
Analyze your own strengths and weaknesses at any time. Expand your strengths at any time. Reduce your weaknesses. Make yourself better, better, more sophisticated.

圖 7. 破窗效應，加強危機意識的認知和控管……
A：發想…… B：目標…… C：願景……
Figure 7. The window breaking effect. Strengthen the awareness and control of crisis awareness...
A: Think... B: Objective... C: Vision...

訂立目標，行動計劃，貫徹實行之，終究會成功的，自然的法則，自然的定律。
Set goals. Action plans. Implement them. They will eventually succeed. The laws of nature. The laws of nature.

圖 8. 碰牆理論，路不轉，人轉，轉念就是力量……
A：發想…… B：目標…… C：願景……
Figure 8. The theory of bumping the wall. The road does not turn. The person turns. The thought is power...
A: Think... B: Objective... C: Vision...

隨時強化專業技能，隨時強化人際溝通能力，這是成功的養分，也是成功的基石。
Strengthen professional skills at any time. Strengthen interpersonal communication skills at any time. This is the nutrient of success. It is also the cornerstone of success.

圖 9. 聚沙成塔理論，累積就是力量……
A：發想…… B：目標…… C：願景……
Figure 9. The theory of gathering sand into a tower. Accumulation is power...
A: Think... B: Objective... C: Vision...

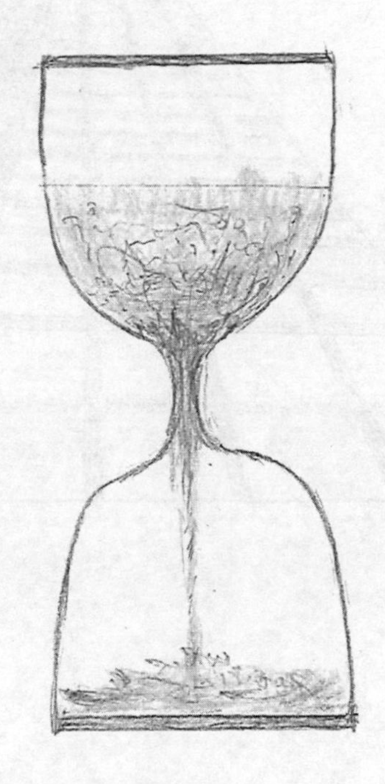

態度＋人際關係是成功的養分，也是成功的保證。
Attitude + interpersonal relationship is the nutrient of success. It is also the guarantee of success.

圖 10. 思維寬廣，目標遠大，大就是力量……
A：發想…… B：目標…… C：願景……
Figure 10. Broad thinking. Great goals. Big is power...
A: Think... B: Objective... C: Vision...

成功旅途上是艱辛的，是辛苦的，是充實的，是快樂的，絕不退縮，絕不放棄，這是成功唯一的心態和信念。
On the journey of success. It is hard. It is hard. It is full. It is happy. Never back down. Never give up.

圖 11. 研讀資訊，好好記錄工作狀況，記錄比記憶重要……
A：發想…… B：目標…… C：願景……
Figure 11. Studying information. Keep a good record of your work situation. Recording is more important than memory...
A: Think... B: Objective... C: Vision...

只要你願意，處處是機會。
As long as you like. Everywhere is opportunity.

圖 12. 常懷感恩心，貴人自然來，感恩就是力量……
A：發想…… B：目標…… C：願景……
Figure 12. Always be grateful. The nobles come naturally. Thanksgiving is power...
A: Think... B: Objective... C: Vision...

人因為有理想，生命才有價值。
People have ideals. Life is valuable.

圖 13. 方向正確，抉擇正確，全力以赴，.一定可以到達成功的彼岸……

A：發想…… B：目標…… C：願景……

Figure 13. The direction is correct. The choice is correct. Go all out. You can definitely reach the other side of success...

A: Think... B: Objective... C: Vision...

成功是靠有準備的，成功是靠有機運的，

成功是靠有責任心的，成功是靠有實力的。

Success depends on preparation. Success depends on luck. Success depends on responsibility. Success depends on strength.

圖 14. 常溝通，常檢討開會，人際關係自然就好……
A：發想…… B：目標…… C：願景……
Figure 14. Regular communication. Regular review meetings. Interpersonal relationships are just fine...
A: Think... B: Objective... C: Vision...

窮則變，轉彎，變則通。
The poor is the way. The turn. The way is the way.

圖 15. 常問候客戶，關心客戶，關心就是力量……
A：發想…… B：目標…… C：願景……
Figure 15. Frequent greetings to Keto. Caring for customers. Caring is power...
A: Think... B: Objective... C: Vision...

圖 16. 看準目標，.設定目標，全力以赴，堅持到底，堅持就是力量……

A：發想…… B：目標…… C：願景……

Figure 16. Look at the goal. Set the goal. Go all out. Persist in the end. Perseverance is the strength...

A: Think... B: Objective... C: Vision...

肢體語言，影響你的一生，也影響別人的一生。

Body language. Affects your life. Also affects the life of others.

圖 17. 常讀書，常學習，學習就是力量……
A：發想…… B：目標…… C：願景……
Figure 17. Regular reading. Regular learning. Learning is power...
A: Think... B: Objective... C: Vision...

山不轉，路轉，路不轉，人轉。
Mountain does not turn. Road turns. Road does not turn. People turn.

圖 18. 讀萬卷書，不如行萬里路，增廣視野，視野就是力量……
A：發想…… B：目標…… C：願景……
Figure 18. Reading Wan Wanshu. It's better to travel a thousand miles. Broaden your horizons. Vision is power...
A: Think... B: Objective... C: Vision...

時時面對人，事，物，隨時要保持感謝的心，感恩的心。
Face people, things, things. Always be thankful. Be grateful.

圖 19. 強化自信心的基石，信心就是力量……
A：發想…… B：目標…… C：願景……
Figure 19. Strengthening self-confidence. The cornerstone of confidence. Confidence is power...
A: Think... B: Objective... C: Vision...

肯定+讚賞是自信的培養土，休息是為了走更遠的路。
Affirmation + appreciation is the cultivation ground of self-confidence. Rest is to go a long way.

圖 20. 飛出去就有希望，離開舒適圈，適應環境，改變自己，改變就是力量……
A：發想…… B：目標…… C：願景……
Figure 20. Flying out. There is hope. Leaving the comfort zone. Adapting to the environment. Changing yourself. Change is power...
A: Think... B: Objective... C: Vision...

失敗，挫折，打擊，羞辱，都是你再一次成長的養分。
Failure, frustration, thrashing, humiliation are all the nutrients you grow up again.

圖 21. 集少成多，聚沙成塔，有小錢才有大錢，人不理財，財就不理人……

A：發想…… B：目標…… C：願景……

Figure 21. Set less and build more. Get together in a sand tower. Have little money. Only have big money. People don't manage money. Finance doesn't care about people...

A: Think... B: Objective... C: Vision...

圖 22. 禮上往來，增進彼此友誼關係，建立好的人際關係，人際關係就是力量……

A：發想…… B：目標…… C：願景……

Figure 22. Ritual exchanges. Enhance mutual friendship. Establish good interpersonal relationships. Interpersonal relationships. It is power...

A: Think... B: Objective... C: Vision...

校長，企業家，領導人，就像花圃的園丁，天天灌溉每一顆種籽，天天照顧每一顆種籽（員工），快快成長，健康壯大。

Principal. Entrepreneur. Leader. Just like the gardener of the flowerbed. Irrigating every seed every day. Taking care of every seed every day（employee）. Growing fast. Healthy and strong.

圖 23. 書中自有黃金屋，常看書，常閱覽報章雜誌，常學習，學習就是力量……
A：發想…… B：目標…… C：願景……
Figure 23. The book has its own golden house. Read books often. Read newspapers and magazines often. Study often. Learning is power...
A: Think... B: Objective... C: Vision...

只要肯上進，肯努力，貧窮也能變富有。
As long as you are willing to make progress, be willing to work hard, and poverty can become rich.

圖 24. 心中常保持學習的態度和習慣，習慣就是力量……
A：發想…… B：目標…… C：願景……
Figure 24. I always keep the attitude and habit of learning in my heart. Habit is power...
A: Think... B: Objective... C: Vision...

只要肯用心，肯下功夫，肯學習，處處是機會。
As long as you work hard. You work hard. You learn. Everywhere is an opportunity.

圖 25. 常常陪孩子，家人，間聊出遊，增廣見聞，建立好的親子和家人關係……

A：發想…… B：目標…… C：願景……

Figure 25. Frequently accompany the children. Family. Chatting between the trips. Widely heard. Established good parent-child and family relationship...

A: Think... B: Objective... C: Vision...

只要你願意，成功大門，永遠為你而開。

As long as you are willing. The door to success. Always open for you.

圖 26. 成功就像播種籽，常去澆水，灌溉施肥，才會拙壯長大，才會有收穫……
A：發想…… B：目標…… C：願景……
Figure 26. Success is like sowing seeds. Frequently watering. Irrigation and fertilization. Will be sturdy. Growing up. Will be harvested...
A: Think... B: Objective... C: Vision...

只要我願意，我的未來前景，前途是很有希望的。
As long as I want. My future prospects. The future is very promising.

圖 27. 勇猛，勇敢，勇氣，向前衝，全力以赴，堅持成功……
A：發想…… B：目標…… C：願景……
Figure 27. Bravery. Bravery. Courage. Rush forward. Go all out. Persevere in success...
A: Think... B: Objective... C: Vision...

只要你願意，泥土也會變黃金。
As long as you want, the soil will become gold.

圖 28. 事態變化快，要有清晰的腦袋，正確的抉擇，才有成功的未來……

A：發想…… B：目標…… C：願景……

Figure 28. Things are changing fast. Aclear head is needed. The right choice. Only to have a successful future...

A: Think... B: Objective... C: Vision...

常常懷著，讚美，感謝，感恩的心。

Often with. Praise. Thanks. Grateful heart.

圖 29. 隨時隨地強化自己，充實自己……
A：發想…… B：目標…… C：願景……
Figure 29. Strengthen yourself anytime, anywhere. Enrich yourself...
A: Think... B: Objective... C: Vision...

改變自己，突破困境，終身學習，終身正向思維，勇往直前。
Change yourself. Break through difficulties. Lifelong learning. Lifelong positive thinking. Go forward bravely.

圖 30. 常問候，常關心，常付出，建立好的人際關係，人際關係就是力量……
A：發想…… B：目標…… C：願景……
Figure 30. Constant greetings, constant concern, constant effort, established good interpersonal relationships, interpersonal relationships, is power...
A: Think... B: Objective... C: Vision...

走出舒適圈，突破舒適圈，改變舒適圈。
Out of the comfort zone. Break through the comfort zone. Change the comfort zone.

圖 31. 有遠見，有夢想，人生才有價值，才有意義……
A：發想…… B：目標…… C：願景……
Figure 31. Visionary. Dream. Life is worth. Only meaningful...
A: Think... B: Objective... C: Vision...

情緒＋態度=成功
Emotion ＋ attitude = success

圖 32. 多運動，養成積極態度，正向的生活心態……
A：發想…… B：目標…… C：願景……
Figure 32. Exercise more. Cultivate a positive attitude。
Apositive attitude towards life...
A: Think... B: Objective... C: Vision...

傾聽+讚美+重視+尊重+敬佩=好的人際關係
Listen + praise + value + respect + admiration = good interpersonal relationship

圖 33. 努力工作，努力賺錢，未雨綢繆，時時要有危機意識……
A：發想…… B：目標…… C：願景……
Figure 33. Work hard. Work hard to make money. Plan ahead. Always have a sense of crisis...
A: Think... B: Objective... C: Vision...

圖 34. 努力學習是進步成功的墊腳石……
A：發想…… B：目標…… C：願景……
Figure 34. Study hard. Stepping stone to success...
A: Think... B: Objective... C: Vision...

要訓練好笑容，要訓練好相處，要訓練好合群。
To train a good smile. To train to get along well. To train a group.

圖 35. 掌握時間，不要白白浪費寶貴的光陰……
A：發想…… B：目標…… C：願景……
Figure 35. Master the time. Do n’t waste precious time in vain...
A: Think... B: Objective... C: Vision...

人脈是致富的跳板，人脈不一定是錢脈，而是要有正確的人脈，才正是錢脈。
The network is a springboard for getting rich. The network is not necessarily the money. It is the right network. It is the money.

圖 36. 要有儲蓄的習慣，儲蓄就是力量……
A：發想…… B：目標…… C：願景……
Figure 36. The habit of saving. Saving is power...
A: Think... B: Objective... C: Vision...

人脈是致勝的關鍵，也是事業成功的保證。
Networking is the key to winning. It is also the guarantee of career success.

圖 37. 發揮想像力，發揮創造力，發揮內在潛力……
A：發想…… B：目標…… C：願景……
Figure 37. Use your imagination. Use your creativity. Use your inner potential.
A: Think... B: Objective... C: Vision...

圖 38. 投資理財，要善用身邊資源，財生財，創造大的邊際效益……

A：發想…… B：目標…… C：願景……

Figure 38. Investment and finance. Make good use of the resources around you. Wealth and wealth. Creating big marginal benefits...

A: Think... B: Objective... C: Vision...

悲觀的人，永遠看到問題背後等惱人的問題。

Pessimistic person. Always see annoying problems behind the problem.

圖 39. 有耐心，有毅力，持續前進，一定會成功……
A：發想…… B：目標…… C：願景……
Figure 39. Have patience. Have perseverance. Continue to move forward. It will be successful...
A: Think... B: Objective... C: Vision...

圖 40. 努力啃書，增長知識，知識就是力量……
A：發想…… B：目標…… C：願景……
Figure 40. Efforts to nibble. Increase knowledge. Knowledge is power...
A: Think... B: Objective... C: Vision...

做人，做事，要做一個有溫度的人，有熱誠的人，有情感的人。
Be a person. Do something. Be a person with temperature. Have a person with enthusiasm.

圖 41. 勤奮學習，吸收新的觀念，好的觀念，觀念就是力量……

A：發想…… B：目標…… C：願景……

Figure 41. Diligent study. Absorption of new ideas. Good ideas. Ideas are power...

A: Think... B: Objective... C: Vision...

有關係=沒關係，沒關係，就是要去找關係，找不到關係，就是要去拉關係，拉不到關係，就是有關係。

It does not matter. It does not matter. It does not matter. It is to find a relationship. It does not find a relationship. It is to pull a relationship.

圖 42. 要有時間管理概念，時間就是力量……
A：發想…… B：目標…… C：願景……
Figure 42. There must be a concept of time management. Time is power...
A: Think... B: Objective... C: Vision...

華人獨特的人際關係：
A.認親戚 B.拉關係 C.攀交情 D.做人情 E.鑽營 F.送禮。
The unique interpersonal relationship of the Chinese:
A. Recognize relatives B. Pull relationships C. Cultivate friendship
D. Do human relations E. Drill camp F. Give gifts.

圖 43. 培養孩子進步的能力和勇氣，能力就是力量……
A：發想…… B：目標…… C：願景……
Figure 43. Cultivating children's ability and courage to enter the market. Ability is strength...
A: Think... B: Objective... C: Vision...

先認識人，再交流，再交心，這是交朋友最佳上策。
Meet people first. Then communicate. Then make friends. This is the best way to make friends.

圖 44. 人要理財，財會理人，人要愛錢，錢會愛人……
A：發想…… B：目標…… C：願景……
Figure 44. People want to manage money. Finance and accounting manage people. People want to love money. Money will love people...
A: Think... B: Objective... C: Vision...

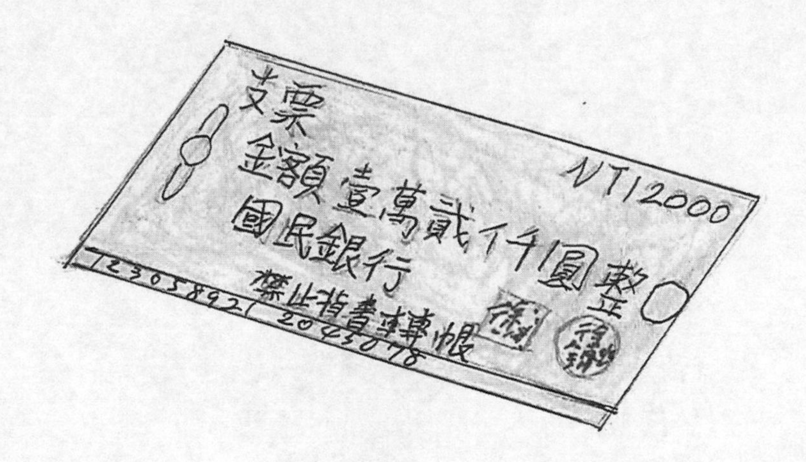

做人，做事，正確比速度更重要。
Being a person. Doing things. Correctness is more important than speed.

圖 45. 投資要分散風險，要有危機意識概念……
A：發想…… B：目標…… C：願景……
Figure 45. Investment should diversify risks.
Aconcept of crisis awareness is needed...
A: Think... B: Objective... C: Vision...

一個人的成功，15%取決於專業技術，85%取決於人際溝通能力。
One's success. 15% depends on professional skills. 85% depends on interpersonal communication skills.

圖 46. 學習是人生的墊腳石，也是人生成功的墊腳石……
A：發想…… B：目標…… C：願景……
Figure 46. Learning is a stepping stone to life. It is also a stepping stone to success in life...
A: Think... B: Objective... C: Vision...

記錄比億更重要，格局決定決局。
Record is more important than 100 million. The pattern determines the ending.

圖 47. 責任愈大成就就愈大，堅持是走向成功的必要條件……
A：發想…… B：目標…… C：願景……
Figure 47. The greater the responsibility. The greater the achievement. Persistence is a necessary condition for success...
A: Think... B: Objective... C: Vision...

公司運作，決策，效益比效率更重要。
Company operation. Decision-making. Benefit is more important than efficiency.

圖 48. 多學習，不恥下問，精益求精，這是學習的基本精神願景……
A：發想…… B：目標…… C：願景……
Figure 48. Learn more. Ask without shame. Keep improving. This is the basic spiritual vision of learning……
A: Think... B: Objective... C: Vision...

好心情，好態度，好人緣，就會有好的人際關係。
Good mood. Good attitude. Good relationship. There will be good relationships.

圖 49. 人生旅途中，方向要正確……
A：發想…… B：目標…… C：願景……
Figure 49. In the journey of life. The direction should be correct.
A: Think... B: Objective... C: Vision...

熱愛你的工作，熱愛你的朋友，熱愛你的生活，這是人生成功要素。
Love your work. Love your friends. Love your life. This is the key to success in life.

圖 50. 常常要寫信問候，請安，關心，增進大家友誼情感……
A：發想…… B：目標…… C：願景……
Figure 50. Always write letters and greetings. Please be safe. Concern. Increase your friendship and friendship...
A: Think... B: Objective... C: Vision...

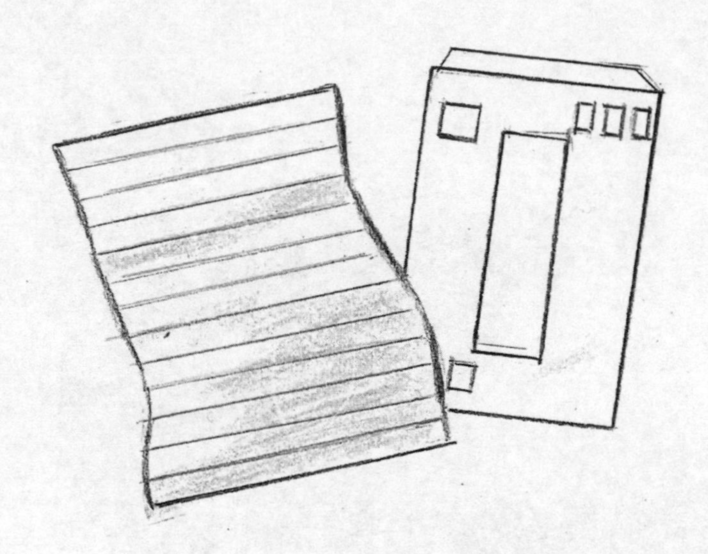

好的人際關係，會帶來好的人生結局。
Good interpersonal relationship will bring good life ending.

圖 51. 掌握最新資訊，投資要謹慎正確……
A：發想…… B：目標…… C：願景……
Figure 51. Master the latest information. Be cautious and correct when investing...
A: Think... B: Objective... C: Vision...

好的溝通能力＋好的人際關係，是事業成功的關鍵所在。
Good communication skills ＋ good interpersonal relationships. It is the key to successful career.

圖 52. 蝴蝶效應，小偏差，會影響未來的大偏差，要注意小地方或細節的影響力……
A：發想…… B：目標…… C：願景……
Figure 52. Butterfly effect. Small deviations. Large deviations that will affect the future. Pay attention to the influence of small places or details...
A: Think... B: Objective... C: Vision...

怎麼樣去發掘和開發自己潛在的能量，相信自己，肯定自己，自己是可以的，這是潛能開發最大的推動力。
How to discover and develop your potential energy. Believe in yourself, affirm yourself, and be yourself. This is the biggest driving force for potential development.

圖 53. 多看書，多閱覽最新資訊，資訊就是力量……
A：發想…… B：目標…… C：願景……
Figure 53. Read more books. Read the latest information. Information is power...
A: Think... B: Objective... C: Vision...

常常給予真誠的讚美和感謝，這是友誼和人際關係昇華的必備元素。
Always give sincere praise and gratitude. This is an essential element for the sublimation of friendship and interpersonal relationships.

圖 54. 同心，同力，合作就是力量……
A：發想…… B：目標…… C：願景……
Figure 54. Concentric. Tongli. Cooperation is power...
A: Think... B: Objective... C: Vision...

常常引發他人心中的渴望，傾聽他人心中的需求，而給予充分的協助和幫助。
Frequently arouse the desires of others. Listen to the needs of others and give them full assistance and help.

圖 55. 凡事要討論，多開會，會激發出好的創意，好的點子……
A：發想…… B：目標…… C：願景……
Figure 55. Everything needs to be discussed. More meetings. Good ideas will be inspired. Good ideas...
A: Think... B: Objective... C: Vision...

圖 56. 播種籽，要多用心照料，就會長出美麗的花朵……
A：發想…… B：目標…… C：願景……
Figure 56. Sowing seeds. Take care of them. Beautiful flowers will grow...
A: Think... B: Objective... C: Vision...

小客戶，要當大客戶來服務，來經營，這是業務工作，成功的基本概念。
Small customers. To serve as big customers. To operate. This is the basic concept of business work. Success.

圖 57. 投資理財，雞蛋不要放在同一個籃子裡，降低投資風險為上策……

A：發想…… B：目標…… C：願景……

Figure 57. Investment and financial management. Do n’ t put eggs in the same basket. Reduce investment risk. For the best strategy...

A: Think... B: Objective... C: Vision...

時時學習，時時增長知識和技能，時時增長智慧。

Learn from time to time. Increase knowledge and skills from time to time. Increase wisdom from time to time.

圖 58. 向前衝，迎向美好未來……
A：發想…… B：目標…… C：願景……
Figure 58. Rush forward. Towards a better future...
A: Think... B: Objective... C: Vision...

愛茵斯坦說，一個人的成功常取決於轉折點上。
Einstein said that one′s success often depends on the turning point.

圖 59. 馬到成功，全力以赴，終身不放棄……
A：發想…… B：目標…… C：願景……
Figure 59. Success in success. Go all out. Don't give up for life...
A: Think... B: Objective... C: Vision...

成功的銷售或服務，就是瞭解並做到超越客戶的期待和期望。
Successful sales or service. It is to understand and exceed customer expectations and expectations.

圖 60. 螞蟻兵團，團結就是力量……
A：發想…… B：目標…… C：願景……
Figure 60. Ant Corps. Unity is strength...
A: Think... B: Objective... C: Vision...

隨時檢視自己的工作效率，想辦法倍增工作效率。
Check your work efficiency at any time. Find ways to increase work efficiency.

C.字樣範例 Sample Word Templat

想想看……想像一下……寫下來……

Think about it... imagine... write it down...

C1：1.很友善的 2.有信心的 3.很誠懇的 4.有親和力的 5.正向的態度 6.負面的態度……

C1: 1. Very friendly 2. Confident. 3 Very sincere. 4 Affinity 5 Positive attitude. 6 Negative attitude...

C2：1.有目標的 2.有力量的 3.有影響力的 4.願意幫助人的 5.願意付出的人 6.有理想的……

C2: 1 Targeted 2. Powerful 3. Influential 4. People willing to help 5. People willing to give 6. Have ideals...

C3：1.笑容的力量 2.溝通的力量 3.專業知識的力量 4.專業技能的力量 5.讚美的力量 6.肢體語言的力量……

C3: 1. The power of smile 2. The power of communication 3. The power of professional knowledge 4. The power of professional skills 5. The power of praise 6. The power of body language...

C4：1.重挫中找機會 2.從機會中找困難 3.時時說對不起的話 4.時時說謝謝的話 5.時時說感恩的話 6.時時與人好相處……

C4: 1. Finding opportunities in times of frustration 2. Finding difficulties from opportunities 3. Saying sorry at all times 4. Saying thanks at all times 5. Saying thanks at all times 6. Getting along with others at all times...

銷售之所以成功，就是能掌握市場的需求，大量的行銷概念。

The reason why sales are successful is to be able to grasp the needs of the market. A large number of marketing concepts.

C5：1.如何聰明的工作……2.如何增進親和力……3.如何增進自信心……4.如何增進溝通力……5.如何增進創新力……6.如何增進領導力……
C5: 1. How to work smartly... 2. How to increase affinity... 3. How to increase self-confidence... 4. How to improve communication... 5. How to increase innovation... 6. How to increase leadership...

C6：1.如何建立好的人際關係 2.常常懷著讚美，感謝，感恩的心……3.做一個有溫度的人，勿做一個冷漠的人 4.做一個有愛心的人，有影響力的人……5.常常鼓勵人家，肯定人家，讓對方覺得受重視，有重要感……6.對人，對事，對物，要抱著熱忱熱情來相待相處，有來有往，建立好的互動人際關係……
C6: 1. How to build a good interpersonal relationship 2. Often with praise. Thanks. Grateful heart... 3. Be a warm person. Do n't be a cold person. 4. Be a caring person. Influential Powerful people... 5. Encourage others. Affirm others. The other party feels valued. Have an important sense... 6. Treat people, things, things, get along with each other with warm enthusiasm. Good interactive relationships...

C7：1.怎麼樣……讓自己更成功……2.決心+毅力+永不放棄=成功 3.細心+耐心+恆心=成功……4.學習+研究+創新=成功 5.合群+合作+團隊力量=成功 6.情緒+態度+感恩=成功……
C7: 1. How to... make yourself more successful... 2. Determination + perseverance + never give up = success 3. Attention + patience + perseverance = success... 4. learning + research + innovation = success 5. group + cooperation + team Strength = Success 6. Emotion + Attitude + Sense = success...

C8：1.走出舒適圈……2.突破舒適圈……3.改變舒適圈……4.挑戰困難，勇往直前……5.挫和失敗是成功的基石……6.不灰心，

隨時掌握最新的資訊，隨時檢討，分析，做決策，納入公司決策一環，並確實實行之。
Keep up to date with the latest information at any time. Review, analyze, make decisions at any time, and incorporate it into the company's decision-making process.

不氣餒，不放棄，大步邁向成功之路……
C8: 1. Go out of the comfort zone... 2. Break through the comfort zone... 3. Change the comfort zone... 4. Challenge the difficulties. Go forward courageously... 5. Frustration and failure are the cornerstones of success... 6. Do n't be discouraged. Do n't be discouraged. Do n't give up. Big Walk towards the road to success...

C9：1.怎麼樣做好情緒管理的主人……2.好的情緒，會帶來好的人際關係……3.好的態度，會帶來好的結果……4.好的人品，會帶來好尊重和敬佩……5.好的學習，會帶來豐富的知識……6.好的習慣，會帶來好的生活……
C9: 1. How to master a good emotional management... 2. Good emotion. Will bring good interpersonal relationship... 3. Good attitude. Will bring good results... 4. Good character. Good respect and admiration... 5. Good learning. Will bring a wealth of knowledge... 6. Good habits. Will bring a good life...

C10：1.笑容+貼心+幫助=成功……2.善良+援助+信任=成功……3.學習+專注+實現=成功……4.交流+正確+結果=成功……5.肯定+趨勢+頂端=成功……6.謙卑+惜緣+惜福=成功……
C10: 1 Smile + Intimate + Help = Success... 2. Kindness + Assistance + Trust = Success... 3. Learning + Care + Achievement = Success... 4. Communication + Correct + Result = Success... 5. Affirmation + Trend + Top = Success... 6. Humility + Cherish Fate + Cherish Blessing = Success...

C11：1.如何培養好笑容……2.如何培養好相處……3.如何培養好合群……4.怎麼樣練習好笑容……5.怎麼樣練習好相處……6.怎麼樣練習好合群……
C11: 1. How to cultivate a good smile... 2. How to cultivate a good

隨時分析自己的優點和缺點，隨時擴大優點，縮減缺點，使自己變得更優質，更卓越，更精進。
Analyze your own strengths and weaknesses at any time. Expand your strengths at any time. Reduce your weaknesses. Make yourself better, better, more sophisticated.

relationship... 3. How to cultivate a good group... 4. How to practice a good smile... 5. How to practice getting along... 6. How to practice getting together...

C12：中國的人際互動有哪些？例如「九同」這些……1.同宗（宗親）…….2.同鄉……3.同學（同窗）……4.同事（同伴）……5.同行……6.同好…… 7 .同胞……8.同年……9.同好……

C12: What are the interpersonal interactions in China? For example, "Nine Tongs"... 1. Tongzong （clan）... 2. Tongyin... 3. Classmates （same classmates）... 4. Colleagues （peers）... 5. Peers... Fellowship... 7. compatriots... 8. same year... 9. fellowship...

C13：中國社會的人際關係有那些……1.同宗（宗親）2.同鄉 3.同學（同窗）4.同事（同伴）5.同行 6.同好（同道.同志）7.同胞 8.同年。

A：發想…… B：目標…… C：願景……

C13: What are the interpersonal relationships in Chinese society... 1 Tongzong（clan）2. Tongxiang 3. Classmates（classmates）4. Colleagues（companion）5. Peer 6. Tonghao（tongdao. Comrade）7. Compatriots 8. Same year.

A: Think... B: Objective... C: Vision...

C14：中國社會的人際關係，例如「五倫」1.君臣 2.父子 3.夫婦 4.兄弟 5.朋友。

C14: Interpersonal relationship in Chinese society. For example, "Five Lun" 1. Monarch 2. Father and son 3. Couple 4. Brother 5. Friends.

C15：如何善用在 1.親密區 2.私交區 3.社交區 4.公眾區等經營你

的人際關係……

C15: How to make good use of your personal relationships in 1. intimate zone 2. private friendship zone 3. social zone 4. public zone etc...

C16：1.建立好的人際關係……2.怎麼樣培養你的熱情……3.如何善用親密區的關係……4.如何善用社交區的關係 5.如何善用公眾區的關係……

A：發想…… B：目標…… C：願景……

C16: Establishing good interpersonal relationships... 2. How to cultivate your enthusiasm... 3. How to make good use of. Intimate zone relationships... 4. How to make good use of social zone relationships 5. How to make good use of public zone relationships...

A: Think... B: Objective... C: Vision...

C17：1.好的人際關係會帶來好的人生結局……2.人際溝通重在傾聽，情感的投射……3.人的相處重在交流，交心，信賴……4.好的人際關係是生活成功的保證……5.好的傾聽習慣是人際關係成功的第一步……

C17: Good interpersonal relationships. Bring good life endings... 2. Interpersonal communication focuses on listening. Emotional projection... 3. People get along with communication. Interpersonal. Trust... 4. Good interpersonal relationships. It is the guarantee of success in life... 5. Good listening habits. It is the first step in the success of interpersonal relationships...

C18：好的溝通能力+好的人際關係，是事業成功的關鍵所在。

C18: Good communication skills + good interpersonal relationship is the key to success in business.

隨時強化專業技能，隨時強化人際溝通能力，這是成功的養分，也是成功的基石。

Strengthen professional skills at any time. Strengthen interpersonal communication skills at any time. This is the nutrient of success. It is also the cornerstone of success.

C19：1.人際關係的影響力……2.人際關係的功能和力量……3.職場人際關係和團隊合作的績效……4.人際關係效益比效率更重要……5.情感+人際關係=溝通模式……6.好心情，好態度，就是好的人際關係……

C19: The influence of interpersonal relationship... 2. The function and power of interpersonal relationship... 3. The cumulative effect of interpersonal relationship and teamwork in the workplace... 4. The benefit of interpersonal relationship is more important than efficiency... 5. Emotion + interpersonal relationship = Communication mode... 6. Good mood, good attitude, good interpersonal relationship...

C20：1.人脈不一定是錢脈，正確的人脈才是錢脈，才是錢潮……2.尊重對方是溝通的開始，是溝通的延續……3.要熱愛自己的工作，熱愛自己的生活，這是成功要素……4.要強烈建立自信心，就是不害羞.不自卑，不退縮的基本概念……5.肯定和讚賞是自信的培養土 6.時時提昇孩子的責任心……

A：發想…… B：目標…… C：願景……

C20: 1.The network is not necessarily the money. The correct network. It is the money. It is the money tide... 2. Respecting the other party is the beginning of communication. It is the continuation of communication... 3. To love your work. Love your own Life. This is the key to success... 4. To build self-confidence strongly. It is not to be shy. No inferiority. The basic concept of not retreating... 5. Affirmation and appreciation are the cultivation grounds of self-confidence. 6. Improving children 's sense of responsibility...

A: Think... B: Objective... C: Vision...

C21：1.成功的人，勇往直前，常記住經驗……2.失敗的人，常退縮不前，忘記教訓和經驗……3.成功的人，懂得創造機會……4.成功的人，懂得掌握機會……5.失敗的人，守株待兔，

態度+人際關係是成功的養分，也是成功的保證。

Attitude + interpersonal relationship is the nutrient of success. It is also the guarantee of success.

錯失機會……6.成功的人，稻穗長得越豐盛，腰彎得越低，謙卑的性格……

A：發想…… B：目標…… C：願景……

C21:.1. Successful people. Go forward courageously. Always remember experience... 2. Failure people. Always retreat. Forget lessons and experience... 3. Successful people. Know how to create opportunities... 4. Successful people. Yes... 5. Failure. Standing by the rabbits. Missing opportunity... 6. Successful. The richer the rice bale grows. The lower the waist bends. The humble character...

A: Think... B: Objective... C: Vision...

C22：1.如何發掘，發展兒童青少年的潛能……2.常常給兒童正向的觀念，多讚美，多鼓勵，多肯定，給予正向的力量……3.常常給予或教導正確的人生觀，正確的人生價值觀……4.常常撥空跟孩子聊聊天，談談孩子生活的想和心事……5.常常傾聽孩子生活趣事，建立好親子互動關係……6.建立好親子互動關係，建立好與青年互動的關係……

A：發想…… B：目標…… C：願景……

C22: 1. How to discover. Develop the potential of children and adolescents... 2. Give children a positive concept often. More praise, more encouragement, more affirmation. Give positive force... 3. Always give or teach a correct outlook on life. The correct values of life... 4. often take the time to chat with the children. Talk about the children 's life thoughts and thoughts... 5. often listen to the children 's life interesting. Establish a good parent-child interaction... 6. Establish a good parent-child interaction. Interactive relationship...

A: Think... B: Objective... C: Vision...

C23：1.成功就是當下抓住機會……2.成功就是大量的行動……3.成功就是專注心思，全力以赴……4.成功就是努力向前，絕不退縮……5.贏在學習，勝在變，輸在猶豫……6.成功就是相信自己，因為相信就是力量……

成功旅途上是艱辛的，是辛苦的，是充實的，是快樂的，絕不退縮，絕不放棄，這是成功唯一的心態和信念。

On the journey of success. It is hard. It is hard. It is full. It is happy. Never back down. Never give up.

A：發想…… B：目標…… C：願景……

C23: 1. Success is to seize the opportunity now... 2. Success is a lot of action... 3. Success is to pay attention. To go all out... 4. Success is to work hard. Never shrink back... 5. Win in learning. Win Changing, losing and hesitating... 6. Success is to believe in yourself. Because. Belief is strength...

A: Think... B: Objective... C: Vision...

C24：1.成功就是成功的人優點+自己的優點.創造獨特的個人成功模式……2.成功就是整合所有的資源，創造 1+1＞2 的概念……3.成功就是借船過河，借力使力，.創造最優的績效……4.成功就是勇往直前，全力以赴……5.成功就是改變你的思維，改變你的價值觀，正向積極的信念……6.成功就是向成功者學習，不斷精進，創造輝煌的人生……

A：發想…… B：目標…… C：願景……

C24: 1. Success is the merit of the successful person + his own merit. Creating a unique personal success model... 2. Success is integrating all resources. Creating the concept of 1 + 1> 2... 3. Success is borrowing a boat to cross the river. Make effort. Create the best cumulative effect... 4. Success is to move forward. Go all out... 5. Success is to change your thinking. Change your values. Positive positive beliefs... 6. Success is to learn from the successful. Keep improving. Create a brilliant life...

A: Think... B: Objective... C: Vision...

C25：1.學習就是力量……2.態度就是力量……3.改變就是力量……4.溝通就是力量……5.善緣就是力量……6.打拼就是力量……

A：發想…… B：目標…… C：願景……

C25: 1. Learning is power... 2. Attitude is power... 3. Change is power... 4. Communication is power... 5. Good luck is power... 6. Hard work is power...

只要你願意，處處是機會。

As long as you like. Everywhere is opportunity.

A: Think... B: Objective... C: Vision...

C26：1.合作就是力量……2.團隊就是力量……3.分享就是力量……4.鼓勵就是力量……5.激勵就是力量……6.信心就是力量……

A：發想…… B：目標…… C：願景……

C26: 1. Cooperation is strength... 2. Team is strength... 3. Sharing is strength... 4. Encouragement is strength... 5. Incentives are strength... 6. Confidence is strength...

A: Think... B: Objective... C: Vision...

C27：1.如何提昇團隊合作力……2.如何提昇情感凝聚力……3.如何提昇培養員工的向心力……4.如何提昇強化主管的領導力 5.如何提昇主管的溝通力……6.如何強化員工和主管的專業技能和專業知識……

A：發想…… B：目標…… C：願景……

C27: 1. How to improve teamwork... 2. How to improve emotional cohesion... 3.How to improve the centripetal force of training employees... 4. How to enhance and strengthen the leadership of supervisors 5. How to improve the communication of supervisors... 6. How to strengthen employees and supervisor Professional skills and professional knowledge...

A: Think... B: Objective... C: Vision...

C28：1.培養企圖心，有鬥志力……2.記住勿滿於現狀，勿滿於小確幸……3.要徹底擺脫慣性思維與保守……4.誠實的問自己.自己在銷售上，有沒有全力以赴……5.如何借銷信輔助工具來幫助自己銷售……6.常常的問自己，我有沒有吸收新的銷售知識和技能……

A：發想…… B：目標…… C：願景……

C28: 1. Cultivation of ambition. Fighting spirit... 2. Remember. Do n't be content with the status quo. Do n't be too lucky... 3. To get rid of inertial thinking and conservativeness... 4. Ask yourself honestly. Sell yourself Have you gone all out... 5. How to borrow sales letter aids to help yourself sell... 6. Ask yourself often. Have I absorbed new sales knowledge and skills...
A: Think... B: Objective... C: Vision...

C29：1.啟動心靈的力量……2.調整心靈的節奏 3.豐富心靈的節奏大 4.給心靈養分，改變你的人生 5.重整心靈的期望 6.感動心靈的力量.感動眾人。
A：發想…… B：目標…… C：願景……
C29: 1. Start the power of the mind... 2. Adjust the rhythm of the mind 3. Enrich the rhythm of the mind 4. Give nourishment to the mind. Change your life 5. Reshape the expectations of the mind 6. Move the power of the mind. Move everyone.
A: Think... B: Objective... C: Vision...

C30：1.怎麼樣才能賺大錢，事業凡是第一.就能賺大錢……2.技術第一……3.服務第一……4.知識第一……5.人際關係第一……6.公共關係第一……
A：發想…… B：目標…… C：願景……
C30: 1. How to make a lot of money. Acareer is always the first. You can make a lot of money... 2. Technology first... 3. Service first... 4. Knowledge first... 5. Interpersonal relationship first... 6. Public relations first One...
A: Think... B: Objective... C: Vision...

C31：1.常常跟自己說，我希望我成為怎麼樣的一個人……2.我想成為一個有愛心的人 3.我想成為可以手心向下，能幫助別人的人 4.我想成為可以幫助人家創業的人 5.我想成為傾聽別人說

成功是靠有準備的，成功是靠有機運的，成功是靠有責任心的，成功是靠有實力的。
Success depends on preparation. Success depends on luck. Success depends on responsibility. Success depends on strength.

話的人 6.我想成為可以跟別人分享的人……

A：發想…… B：目標…… C：願景……

C31: 1. I often tell myself. What kind of person do I want me to be... 2. I want to be a loving person 3. I want to be able to palm down. Can help others 4. I want to be able to help People who start their own businesses 5. I want to be someone who listens to others 6. I want to be someone who can share with others...

A: Think... B: Objective... C: Vision...

C32：1.常常跟自己說，我的公司能帶給員或人們有什麼助益的……2.公司希望能帶給員工好的生活技能 3.公司希望帶給員工好的生活環境 4.公司希望帶給社會好的氛圍 5.常常想我能為公司做更多的好的建議，常常想，我能為公司做更多事，或更多創新的好點子……

A：發想…… B：目標…… C：願景……

C32: 1. I often tell myself. What can my company bring to employees or people... 2. The company hopes to bring good life skills to employees 3. The company wants to bring good living conditions to employees 4. The company I hope to bring a good atmosphere to society 5. I often think I can make more good suggestions for the company. I often think. I can do more things for the company.

A: Think... B: Objective... C: Vision...

C33：1.常常跟自己說，我對家庭做那些關心和幫助……2.常常想，我對兄弟姊妹有沒有常關心，或幫助或協助兄弟姊妹……3.有沒有常常關心子女，聊天或傾聽子女生活趣事……4.常常有沒有對職場的伙伴關心或幫助……5.有沒有常常關心父母，或老人家的生活狀況……6.有沒有常常對朋友關心，或問好幫助等等……

A：發想…… B：目標…… C：願景……

C33: 1. I often say to myself. I care about and help the family…… 2. I

often think. I care about my brothers and sisters. Do I always care or help or assist my brothers and sisters. 3. I often care about my children. Chat or Listen to children 's life anecdotes... 4. Do you often care or help your partners in the workplace... 5. Do you often care about your parents. Or the living conditions of the elderly... 6. Do you often care about your friends. Or ask for help, etc...
A: Think... B: Objective... C: Vision...

C34：1.透過人際關係訓練，你收獲最多的是什麼，或有那些值得驕傲的……2.要時時的問自己說，我在人際關係上做了那些的努力……3.常付出，好相處，是人際關係的成功關鍵所在……4.好的人際關係，是生意事業興隆的培養土 5.業務的興隆，好的人際係，是決定你業務和事業在未來有無發展的重要命脈……6.人潮就是錢潮.人脈就是錢脈……
C34: 1. Through interpersonal relationship training. What do you gain the most? Or are you proud of? 2. Ask yourself from time to time. I am in interpersonal relationships. I have made those efforts... 3. Give often. Getting along well is the key to the success of interpersonal relationships... 4. Good interpersonal relationships is the breeding ground for business and career prosperity. 5. Business is prosperous. Good interpersonal relationships are important lifelines that determine whether your business and career will develop in the future …… 6. The crowd is the money flow. The network is the money flow……

C35：1.透過潛能開發訓練，你收獲最多的是什麼，.或對你幫助有那些……2.相信它，鼓勵它，時時發揮自己潛在的能量 3.發揮潛在能量，是人類成功最重要的法寶 4.潛能是需要透過學習，練習，培養正能量，或渴望的正能量，才能被激發出來的……5.人類的潛在能量，只有用到 10%，90%是需要等你去開發的……6.潛在能量的開發，是人類最重要，最寶貴的無形資產……

走過的路，走過的經驗，就是你的資產。
The road traveled. The experience traveled is your asset.

C35: 1. Development training through potentials. What do you gain the most. Or help you... 2. Believe it. Encourage it. Use your potential energy from time to time. 3. Use potential energy. It is the most important for human success The magic weapon. 4. Potential needs to be stimulated by learning, practicing, cultivating positive energy, or craving positive energy... 5. Human potential energy. Only 10% is used. 90% is waiting for you to develop... 6. The development of potential energy. It is the most important human being. The most precious intangible asset...

C36：1.透過行銷開發訓練，你收獲最多的是什麼，或對你銷售有那些幫助……2.推銷動作，行銷動作，是銷售或成功的關鍵所在……3.相信自己，肯定自己，自己一定可以成為最優秀的推銷員，最棒的行銷員……4.推銷.就是要不斷持續的建立好人際關係，人際互動系統……5.友善的推銷.友善的行銷，會帶來巨大的財富和快樂……6.推銷和行銷，要用友善的態度來推銷和行銷.這是一件有價值，最快樂的事……

A：發想…… B：目標…… C：願景……

C36: 1. Training through marketing development. What do you gain the most. Or help you in sales... 2. Marketing actions. Marketing actions. The key to sales or success... 3. Believe in yourself. Be sure of yourself. Can become the best salesman. The best salesman... 4. Markcting. Is to continuc to cstablish good intcrpcrsonal rclationships. Interpersonal interaction system... 5. Friendly sales. Friendly marketing. Will bring great wealth and happiness …… 6. Selling and marketing. It is necessary to promote and market with a friendly attitude. This is a valuable thing. The happiest thing……

Λ: Think... B: Objective... C: Vision...

C37：1.透過市場資訊的訓練，你收獲是什麼，或對市場的認知有那些……2.要常常對市場環境的瞭解，市場的變化等等……3.

市場是一個大環境，你必須常常去關心，去做市場變化分析……4.國內市場資訊，國際市場資訊，要隨時隨地去掌握最新的資訊……5.市場資訊蒐集，可增加你做未來做投資決策能力…….6.市場好與不好，關係著你生意成敗的關鍵所在……

C37: 1. Training through market information. What do you gain. Or what do you know about the market... 2. Always have an understanding of the market environment. Changes in the market, etc. 3. The market is a big environment. You must often To care. To do market analysis... 4. Domestic market information. International market information. Keep up to date with the latest information anytime, anywhere... 5. Market information collection. Can increase your ability to make future investment decisions... 6. The market is good or not. The key to your business success or failure...

C38：1.透過市場開發訓練，你收獲最多的是什麼，或對市場瞭解和分析有那些……2.產品要時常推陳出新.贏得先機……3.產品價位的決策是決定產品的未來發展……4.產品+宣傳=市場行銷成功……5.宣傳+廣告=決定產品銷售成敗……6.產品好+宣傳+廣告=市場銷售成功……

C38: 1. Through market development training. What do you gain the most? Or what do you know and analyze about the market... 2. Products should be constantly updated. Win opportunities... 3. The decision of product price is to determine the future development of the product... .Product + promotion = marketing success... 5. Promotion + advertisement = determine the success or failure of product sales... 6. Good product + promotion + advertisement = marketing success...

C39：1.透過產品技術訓練，你收獲最多的有那些，對你未來幫助最多的有那些……2.產品要多元化，隨時隨地要去瞭解人們的需求是什麼……3.產品技術，隨時隨地要強化精進……4.產品

山不轉，路轉，路不轉，人轉。

Mountain does not turn. Road turns. Road does not turn. People turn.

知識，隨時隨地要強化精進……5.銷售技巧，隨時隨地要強化精進……6.人際關係，隨時隨地要強化精進……

C39: 1. Through product technology training. The ones you gain the most from. The ones that will help you the most in the future... 2. The products should be diversified. You need to know anytime, anywhere. Strengthen intensiveness anywhere... 4. Product knowledge. Strengthen intensiveness anytime, anywhere... 5. Sales skills. Strengthen intensiveness anytime, anywhere... 6. Interpersonal relationships. Strengthen intensiveness at any time...

C40：1.透過經營策略的訓練，你收獲最多的有什麼，或對你決策判斷有那些……2.好的經營計劃，真的很重要……3.好的經營策略，真的很重要……4.好的經營規劃，好的經營企劃，真的很重要……5.學習別人好的計劃，好的企劃，好的策略，真的很重要……6.學習別人，好的思維，好的創意，好想望，真的很重要……

A：發想…… B：目標…… C：願景……

C40: 1. Through business strategy training. What do you gain the most? Or what are your judgments for your decision... 2. Good business plan. Really important... 3. Good business strategy. Really important... 4 .Good business planning. Good business planning. It's really important... 5. Learn from others' good plans. Good planning. Good strategy. Really important... 6. Learning from others. Good thinking. Good creativity. Good hope. It's really important...

A: Think... B: Objective... C: Vision...

C41：1.產品轉介紹的重要性，你收獲最多有那些，帶來那些巨大效益……2.客戶轉介紹，真的很重要……3.隨時隨地要請託客戶，幫我們轉介紹，真的很重要……4.產品銷售績效和轉介紹成比……5.要請託客戶，幫我們推薦優質客戶，真的很重要……

A：發想…… B：目標…… C：願景……

C41: 1. The importance of product referrals. The ones you gain the most. Bring those huge benefits... 2. Customer referrals. Really important... 3. Ask clients anytime, anywhere. Help us refer referrals. Really Important... 4. Product sales effectiveness and referral ratio... 5. Please entrust customers. Help us recommend high-quality customers. Really important...

A: Think... B: Objective... C: Vision...

C42：1.複雜的事情簡單做，簡單重複做每一件事情，你就成功了……2.學習要透過複習，練習，才有出息，不複習，不練習，就沒有出息……3.創業很有潛力，需要有很好的學習力……4.成功不一定要贏在起跑點上，但要一定贏在轉捩上……5.怎樣翻轉你的命運，選對事業方向，勇往直前.始終不放棄……6.經營你的事業，要像經營你的人生一樣，用力，用心，全力以赴，堅持成功……

A：發想…… B：目標…… C：願景……

C42: 1. Complicated things are simple to do. Simple and repeat to do everything. You will be successful... 2. Learning must be through review. Practice. Only rewards. Without review. Without practice. There is potential. You need to have a good learning ability... 4. Success does not necessarily have to win at the starting point. But you must win at the transition... 5. How to flip your destiny. Choose the right career direction. Go forward. …… 6. Run your business. Be like running your life. Force. Diligently. Go all out. Persevere in success……

A: Think... B: Objective... C: Vision...

C42：1.一個人成功，要有堅定的信念……2.目標+計劃 x 行動=成功計劃……3.人生最重要的不是奮鬥，而是你做出最正確的抉擇……4.成功是小收獲+小收獲+小收獲=成功……5.成功是小成就+小成就+小成就=成功一 6.堅持信念，目標正確，勇往直前，終身堅持，永不放棄，你就可以成功……

A：發想…… B：目標…… C：願景……

C42: 1.Aperson's success. Have a firm belief... 2. Goal + plan x action = successful plan... 3. The most important thing in life is not struggle. It is you who make the most correct choice... 4. Success is a small gain + Small gains + small gains = success... 5. Success is small achievements + small achievements + small achievements = success 1 6. Adhere to faith. Goals are right. Go forward courageously. Persevere for life. Never give up. You can succeed...

A: Think... B: Objective... C: Vision...

C43：1.設定目標，一定要設定期限，要行動，否則豪無意義……2.你的方向有多大，你的希望就有多大，你的成就就多大……3.時間要切割分段，來實現，這樣就比較容易達成目標……4.想達成目標，要借重或尋找達成目標的輔助工具來達標，這是很重要的過程……5.目標的設定，方向要正確縝密，期限要清晰明確，全力以赴，方可成功……6.目標正確，方向正確，行動就對了，不猶豫，不往後看，行動就對了……

A：發想…… B：目標…… C：願景……

C43: 1. Set goals. Be sure to set deadlines. Take action. Otherwise, ho is meaningless... 2. How big is your direction. How big is your hope. How big is your achievement... 3. Time to cut the segment. Come Achieve this. It is easier to achieve the goal... 4. Want to achieve the goal. It is necessary to borrow the weight or look for the auxiliary tool to achieve the goal. This is an important process... 5. The setting of the goal. The direction should be correct and meticulous. The deadline To be clear and clear. Go all out. Only to succeed... 6. The goal is correct. The direction is correct. The action is right. Do n't hesitate. Do n't look back. The action is right...

A: Think... B: Objective... C: Vision...

失敗，挫折，打擊，羞辱，都是你再一次成長的養分。

Failure, frustration, thrashing, humiliation are all the nutrients you grow up again.

C44：1.想法有多寬多大，市場就有多寬多大……2.給自己一個成功的時間表……3.你肩上扛有多大責任，你的成就就有多大……4.只要學習，不去練習，不去複習，是不會有出息的……5.人生最重要的不是努力，而是做出最正確的抉擇……6.學習與改變，成長與成功，是你人生勝敗的關鍵所在……

A：發想…… B：目標…… C：願景……

C44: 1. How wide and how big the idea is. How wide and how big the market... 2. Give yourself a timetable for success... 3. How much responsibility do you carry on your shoulders. How big is your achievement... 4. Just study. No To practice. Not to review. There will be no good... 5. The most important thing in life is not hard work. It is to make the most correct choice... 6. Learning and change. Growth and success. The key to your success or failure...

A: Think... B: Objective... C: Vision...

C45：1.想法的寬與遠，和你的事業成就成正比……2.正能量，想望，與你的事業成正比……3.好的態度和心態，與你的事業成正比……4.好的人際關係，與你的事業成正比……5.消極心態與態度，與你的事業成負比……6.積極奮發的態度和心態，與你的事業成正比……

A：發想…… B：目標…… C：願景……

C45: 1. The breadth and distance of ideas. Proportionally proportional to your career achievement... 2. Positive energy. Expectation. Proportionally proportional to your career... 3. Good attitude and mentality. Your interpersonal relationship is directly proportional to your career... 5. Negative mentality and attitude. Negatively proportional to your career... 6. Positive attitude and mentality are directly proportional to your career...

A：發想…… B：目標…… C：願景……

A: Think... B: Objective... C: Vision...

工作的開始，正是學習和磨練的開始，隨時保持成長進步和超越。

The beginning of work. It is the beginning of learning and discipline. Keep growing and progressing at any time.

C46：1.成功的經驗，可以分享更多的經驗 2.小成功靠自己……3.大成功靠團隊……4.永恆的成功靠系統化……5.失敗的經驗，可以避免更多的失敗……6.失敗是成功的培養土……

A：發想…… B：目標…… C：願景……

C46: 1. Successful experience. You can share more experience. 2. Small success depends on yourself... 3. Big success depends on the team... 4. Eternal success depends on systematization... 5. Failure experience. You can avoid more Failure... 6. Failure is the breeding ground for success...

A: Think... B: Objective... C: Vision...

C47：1.怎麼樣才能變的有錢一點，首先心中要有富中之富的概念……2.人最可悲的是……心中沒有財富的觀念……3.人理財，財就不理人……4.心中要建立人可以窮，心不能窮的信念……5.常常多讀理財書籍，閱覽或多看理財新知……6.財富是靠拼來的，靠勤儉來的，善用資源來的……

A：發想…… B：目標…… C：願景……

C47: 1. How can one become rich. First of all, there must be the concept of wealth in the heart... 2. The saddest thing about a person is that there is no concept of wealth in his heart... 3. People managing money. Wealth ignores people... 4. I want to build a belief that peoplc can be poor. My heart cannot be poor... 5. Read financial management books frequently. Read or read new financial management knowledge... 6. Wealth comes from fighting. Yes. Make good use of resources...

A: Think... B: Objective... C: Vision...

C48：1.如何提升組織的向心力……2.如何提升團隊的凝聚力……3.如何提升個人向心力……4.強化團隊合作合作力……5.強化團隊合作的經驗力……6.強化團隊合作的溝通力……

A：發想…… B：目標…… C：願景……

C48: 1. How to improve the centripetal force of the organization... 2.

How to improve the cohesive force of the team... 3. How to improve the individual centripetal force... 4. Strengthen the teamwork cooperation... 5. Strengthen the teamwork experience... 6. Strengthen the team The communication skills of cooperation...
A: Think... B: Objective... C: Vision...

C49：1.好的思考.好的行動力，是你未來事業最強的動力，最強的力量……2.敢夢，敢想，創造你人生無限的可能……3.信念夠強.堅持到底，成功就在你眼前，這樣不成功也難……4.造就別人，才能成就自己……5.你要時常對自己說，組織網大，商機就大，成功就大……6.有機會設定目標，改變，改變真的很重要……
A：發想…… B：目標…… C：願景……
C49: 1. Good thinking, good action, is the strongest motivation for your future career. The strongest power.... 2. Dare to dream. Dare to think. Create unlimited possibilities in your life... 3. Faith is strong enough. Persevere. Success It's right in front of you. It's hard to be unsuccessful... 4. Make others. You can achieve yourself... 5. You have to say to yourself from time to time. The organization network is big. The business opportunity is big. The success is big... Change. Change is really important...
A: Think... B: Objective... C: Vision...

C50：1.什麼是最好的經營管理，走動式的人力資源管理……2.生產力，最……3.團隊合作力，協調力，是強化生產力的基石……4.掌握最新資讯.掌握人力資源，是企業生產力最佳的培養土……5.企業是人與人的事業，所以必須先做人，再做事的概念……6.人是生產力和企業的命脈……
A：發想…… B：目標…… C：願景……
C50: 1. What is the best management. Walking human resource management... 2. Productivity. Most... 3. Teamwork. Coordination. It is

the cornerstone of strengthening productivity... 4. Master the latest information. Master the human resources. It is the best cultivation ground for enterprise productivity... 5. The enterprise is the business of people and people. So you must be a person first. Productivity and the lifeblood of an enterprise...
A: Think... B: Objective... C: Vision...

C51：1.如何建立龐大的人際組織網……2.如何建立大量的客戶群和組織網……3.組織網建立，0 到 1 比較困難，1 到 10 就比較容易，10 到 100 就更容易了……4.如何建立龐大事業組織網……5.建構龐大的組織網，這是企業永續經營的命脈……6.建立龐大的企業組織網，這是企業成功的法寶……
A：發想…… B：目標…… C：願景……
C51: 1. How to build a huge interpersonal organization network... 2. How to build a large number of customer groups and organization networks... 3. Organization network establishment. 0 to 1 is more difficult. 1 to 10 is easier. 10 to 100 is more It's easy... 4. How to build a huge network of business organizations... 5. Build a huge network of organizations. This is the lifeblood of an enterprise's sustainable operation... 6. Establish a huge network of corporate organizations. This is a magic weapon for business success...
A: Think... B: Objective... C: Vision...

C52：1.如何建立龐大企業組織網……2.如何建立龐大生產組織網……3.如何建立龐大產品組織網……4.如何建立龐大產品銷售網……
A：發想…… B：目標…… C：願景……
C52: 1. How to build a huge enterprise organization network... 2. How to build a huge production organization network... 3. How to build a huge product organization network... 4. How to build a huge product sales network...
A: Think... B: Objective... C: Vision...

只要肯用心，肯下功夫，肯學習，處處是機會。
As long as you work hard. You work hard. You learn. Everywhere is an opportunity.

C53：1.建立龐大的銷售服務組織網……2.服務是銷售的開始……3.服務是銷延續……4.銷售+服務=成功銷售……5.服務是企業銷售.行銷的命脈……6.服務是企業銷售.行銷的法寶……
A：發想…… B：目標…… C：願景……
C53: 1. Establishing a huge sales service organization network... 2. Service is the beginning of sales... 3. Service is the continuation of sales... 4. Sales + service = successful sales... 5. Service is the sales life of an enterprise. The lifeblood of marketing... 6 .Services are corporate sales。
Amagic weapon for marketing...
A: Think... B: Objective... C: Vision...

C54：1.一份事業，一個機會，一個決定，將改變你的一生……2.人生不一定要贏在起跑點上，重要的是要贏在轉折點上……3.學習才有力量，改變真的很重要……4.培養企圖心，戰鬥力，要不斷突破自己的極限……5.激發潛力，激發學習力……6.造就別人，才能成就自己……
A：發想…… B：目標…… C：願景……
C54: 1.Acareer. An opportunity. Adecision. Will change your life... 2. Life does not have to win at the starting point. The important thing is to win at the turning point. 3. Learning is the power . Change is really important... 4. Cultivate ambition. Fighting power. We must constantly break through our limits... 5. Inspire potential. Inspire learning power... 6. Bring up others. In order to achieve oneself...
A: Think... B: Objective... C: Vision...

C55：1.所有的生意，所有的事業，都是人與人的事業，所以經營人際關係很重要……2.隨時隨地要持續提供有價值的產品和訊息給消費者……3.產品和消費者，是企業的命脈……4.真誠告訴自己說，組織夠大，商機就大，企業發展就大……5.企業要常常做員工專業培訓工作，強化員工向心力 6.企業盛衰，是全體員工生活，生存的保障……

A：發想…… B：目標…… C：願景……

C55: 1. All businesses. All businesses. People-to-people businesses. So managing interpersonal relationships is very important... 2. To provide valuable products and information to consumers at any time and anywhere... 3. Products and consumers . Is the lifeblood of the enterprise... 4. Tell yourself sincerely. The organization is big enough. The business opportunities are big. The development of the enterprise is big... 5. The enterprise should often do employee training. Strengthen the centripetal force of the staff. .Guarantee of Survival...

A: Think... B: Objective... C: Vision...

C56：1.銷售讓你更成功，生活更美好 2.銷售是人和人的事業，銷售自然成功……3.銷售是透過學習，練習，複習，生銷售的力量……4.銷售時時要強化銷售知識和銷售技巧……5.銷售，能帶來豐沛的財富收入……6.銷售，靠推銷和行銷，將為你帶來不可能的銷售奇蹟……

A：發想…… B：目標…… C：願景……

C56: 1. Sales make you more successful. Life is better 2. Sales are the business of people and people. Sales are naturally successful... 3. Sales are through learning. Practice. Review. The power of sales... Sales knowledge and sales skills... 5. Sales. Can bring a wealth of wealth income... 6. Sales. By sales and marketing. It will bring you an impossible sales miracle...

A: Think... B: Objective... C: Vision...

C57：1.一個希望，一個追尋，一個執著，成就了明天的你……2.抓住一個機會，抓住一個緣分，抓住一個正確的方向，抓住一個正確的路，將使你事業更輝煌……3.因為有希望，有機會，成功將指日可待……4.抓住每一個緣分，將使你人脈日日增廣，受人歡迎……5，正確的路，要勇往直前，堅持到一分一

只要我願意，我的未來前景，前途是很有希望的。

As long as I want. My future prospects. The future is very promising.

秒……6 因為有希望，有未來，將使你的人生變的不一樣，不平凡……

A：發想…… B：目標…… C：願景……

C57: 1.Ahope. Apursuit. Aperseverance. You who made tomorrow... 2. Hold a chance. Seize a fate. Seize a correct direction. Seize a correct way. The career is more brilliant... 3. Because there is hope. There is a chance. Success will be just around the corner... 4. Seize every one. You will increase your network every day. Popular... 5. The right way. Go forward courageously. Persevere until one minute and one second... 6 Because there is hope. There is a future. It will make your life different. Not ordinary...

A: Think... B: Objective... C: Vision...

C58：1.徹底檢討過往，計劃未來，付出行動，達成目標……2.反省就是力量……3.肯奮鬥，就是力量 4.不怕從來，只怕沒有未來……5.全力以赴，勇往直前，堅持到底，邁向成功之路……6.感動，有愛，有夢想，就是力量……

A：發想…… B：目標…… C：願景……

C58: 1. Thoroughly review the past. Plan for the future. Pay for action. Achieve goals... 2. Introspection is power... 3. Struggle. Strength. 4. Not afraid of never. Just afraid of no future... 5. Go all out. Hold on bravely. Stick to the end. Toward the road to success... 6. Moved. There is love. There are dreams. It is power...

A: Think... B: Objective... C: Vision...

C59：1.學習就是力量……2.夢想就是力量……3.敢夢，敢想，就是力量 4.相信就是力量……5.團隊經驗分享，就是力量……6.團隊合作分享，就是力量……

A：發想…… B：目標…… C：願景……

C59: 1. Learning is power... 2. Dream is power... 3. Dare to dream. Dare to think. Is power. 4. Believe is power... 5. Team experience sharing... Is power. 6. Teamwork sharing. Is power...

A: Think... B: Objective... C: Vision...

C60：1.人際關係就是力量……2.好人緣就是力量 3.感動就是力量 4.合作就是力量 5.堅持就是力量 6.付出就是力量 7.愛心就是力量……
A：發想…… B：目標…… C：願景……
C60: 1. Interpersonal relationship is power... 2. Good people are power 3. Touch is power 4. Cooperation is power 5. Perseverance is power 6. Pay is power. 7. Love is power...
A: Think... B: Objective... C: Vision...

C61：1.歡迎參加井易免店面飲水機，淨水機租用加盟事業團隊……2.井易淨水機事業團隊，是你最佳的學習平台，也是你最佳的學習舞台……3.輕鬆創業，百萬年薪不是夢……4.有意加盟者.經營潛能開發和第二專長訓練……5 每週六舉辦創業說明會……6.專人專屬為您服務，致富就在眼前，成功就在眼前，優質井易企業，期待和您一起共創美好未來。
C61: 1. Welcome to join the Jingyi free store water dispenser. The water purifier rental joins the business team... 2. The Jingyi water purifier business team is your best learning platform. It is also your best learning stage... 3 .Songsong Entrepreneurship. Million annual salary is not a dream... 4. Interested franchisees. Business potential development and second long-term training... 5 Startup briefing every Saturday... 6.Success is at hand. High-quality Jingyi Enterprise. Looking forward to working with you to create a better future.

附錄一：超級手冊
Super Manual

E1.超級創業記録手冊。
A：發想…… B：目標…… C：願景……
E1. Super entrepreneurship record manual.
A: Think... B: Objective... C: Vision...

E2.超級創業目標手冊。
A：發想…… B：目標…… C：願景……
E2. Super Startup Goal Manual.
A: Think... B: Objective... C: Vision...

E3.超級創業成功手冊。
A：發想…… B：目標…… C：願景……
E3. Super Startup Success Handbook.
A: Think... B: Objective... C: Vision...

E4.超級會議手冊。
A：發想…… B：目標…… C：願景……
E4. Super Conference Manual.
A: Think... B: Objective... C: Vision...

E5.超級快速成功手冊。
A：發想…… B：目標…… C：願景……
E5. Super fast success manual.
A: Think... B: Objective... C: Vision...

E6.超級挖掘心靈潛能手冊。
A：發想…… B：目標…… C：願景……
E6. Super Mining Potential Handbook.
A: Think... B: Objective... C: Vision...

E7.超級優勢銷售手冊。
A：發想…… B：目標…… C：願景……
E7. Super Advantage Sales Manual.
A: Think... B: Objective... C: Vision...

E8.超級事業發展手冊。
A：發想…… B：目標…… C：願景……
E8. Super Business Development Manual.
A: Think... B: Objective... C: Vision...

E9.超級自我訓練手冊。
A：發想…… B：目標…… C：願景……
E9. Super self-training manual.
A: Think... B: Objective... C: Vision...

E10.超級公共關係手冊。
A：發想…… B：目標…… C：願景……
E10. Super Public Relations Manual.
A: Think... B: Objective... C: Vision...

E11.超級成敗檢核項目手冊。
A：發想…… B：目標…… C：願景……
E11. Super success and failure inspection project manual.
A: Think... B: Objective... C: Vision...

E12.超級創意激發手冊。

A：發想…… B：目標…… C：願景……

E12. Super creative inspiration manual.

A: Think... B: Objective... C: Vision...

E13.超級成功習慣手冊。

A：發想…… B：目標…… C：願景……

E13. Super success habit manual.

A: Think... B: Objective... C: Vision...

E14.超級個人成功手冊。

A：發想…… B：目標…… C：願景……

E14. Super personal success manual.

A: Think... B: Objective... C: Vision...

E15.超級個人心靈潛能手冊。

A：發想…… B：目標…… C：願景……

E15. Super Personal Mind Potential Manual.

A: Think... B: Objective... C: Vision...

E16.超級人際關係手冊。

A：發想…… B：目標…… C：願景……

E16. Super Interpersonal Handbook.

A: Think... B: Objective... C: Vision...

情緒＋態度=成功

Emotion + attitude = success

附錄二：英文單字索引
English Vocabulary Index

1	Feel	感覺
2	Felt	感受
3	Chamber	會場
4	Progress	前進發展
5	Vitality	生活、生命力
6	Honestly	誠實
7	Grow	成長
8	Culture	文化
9	Sincere	真誠
10	Impart	傳授
11	technique	技術
12	Enterprise	企業
13	Harmonious	和睦
14	Innovate	革新、改革
15	Council	會議
16	Optimal	最理想
17	Exclaim	呼喊
18	Exclusive	專有
19	exercise	習題

20	exert	努力、運用
21	Exhibition	陳列、博覽會
22	exile	放逐
23	Positive	積極
24	Mental	心理、精神
25	Attitude	態度
26	Full	充分
27	develop	發展
28	Potency	潛能
29	what	什麼
30	Why	為什麼
31	Who	誰
32	When	什麼時候
33	Where	什麼地方
34	How	如何做、方法是什麼
35	Brave	勇氣
36	impossible	不可能
37	Try again	嘗試
38	action	行動
39	power	力量
40	let	容許
41	leauing	趨勢
42	detail	細節
43	lie	說謊

合諧就是力量，合作就是力量，團隊就是力量。
Harmony is power. Cooperation is power. Team is power.

44	Liar	說謊
45	Liberate	解放
46	Enlighten	啟發
47	Line	路線
48	Vision	遠景
49	Literature	文學、著作
50	Living	生活、生計
51	Lenliven	使愉快
52	Locality	場地
53	Long	渴望
54	Lordly	貴族
55	Be Loved	愛人
56	Luxurious	奢侈
57	Main	主要
58	Manner	方法
59	Mature	成熟
60	Mate	配偶、夥伴
61	mateer	重要、理由
62	Meaningless	無意義
63	Meadiate	沉思、企圖
64	Melanancholy	沮喪
65	Melody	和諧
66	Member	會員
67	Melk	融化

要訓練好笑容，要訓練好相處，要訓練好合群。

To train a good smile. To train to get along well. To train a group.

68	Menacemeds	威脅
69	Mend	修理、改善
70	Mental	心理
71	Mention	敘述
72	Merit	優點
73	Method	計劃
74	Mileage	旅費
75	Millionaire	百萬富翁
76	Mind	留心
77	Miraculous	神奇的
78	Mission	使命
79	Mode	方法
80	stupid	笨
81	Negative	消極的
82	Negotiation	談判、交涉
83	Nest	巢
84	Enrich	使富足
85	Stratagem	策略
86	Enquiry	調查
87	Enterprising	有冒險心
88	Entertainment	款待
89	Essay	論文
90	Enthusiast	熱心
91	Estate	財產

人脈是致富的跳板，人脈不一定是錢脈，而是要有正確的人脈，才正是錢脈。

The network is a springboard for getting rich. The network is not necessarily the money. It is the right network. It is the money.

92	Eve	前夕
93	Income	收入
94	Event	結果事件
95	Evil	邪惡
96	Excellence	卓越
97	Exchange	交易
98	Excitedly	興奮
99	Exclusion	拒絕
100	Execute	實施、完成
101	Exertion	奮發
102	Exhibiti	陳列、展覽
103	Laugh	嘲笑
104	Lcadership	領袖
105	Keague	聯盟
106	Society	聯盟
107	Lecture	講演
108	Lend	出租
109	Let	容許
110	Liability	責任
111	Schedule	行程
112	Chart	表
113	Analysis chart	營業分析表
114	Weak	虛弱
115	Wise	有智慧

人脈是致勝的關鍵，也是事業成功的保證。

Networking is the key to winning. It is also the guarantee of career success.

116	Writer	作者
117	Traud	欺騙
118	Investment	投資
119	Squad	隊
120	Punish	處罰
121	Receive	收到
122	Repeat	重複
123	Save	存
124	Send	寄
125	Shout	喊叫
126	Show	展示
127	Spend	花費
128	Stick	貼
129	Test	測試
130	Taste	嚐
131	Teach	教導
132	Throw	丟
133	Turn	轉
134	Type	打字
135	Want	想要
136	Wash	清洗
137	Waste	浪費
138	Shall	應該
139	Often	經常

樂觀的人，永遠看到問題背後的機會。
Optimistic person. Always see the opportunity behind the problem.

140	Until	直到
141	Behind	在……之後
142	Below	在……下面
143	Beside	在……旁邊
144	By	藉由
145	For	為了
146	In	在裡面
147	On	在上面
148	Over	越過
149	though	通過
150	Under	在……之下
151	With	與
152	Without	沒有
153	Which	哪一個
154	Each	每個
155	Both	兩者都
156	Most	大多數
157	Everything	一切事物
158	None	沒有（人）
159	Classmate	同學
160	Neighbor	鄰居
161	Malam	女士
162	Joy	樂趣
163	Enjoy	欣賞

悲觀的人，永遠看到問題背後等惱人的問題。
Pessimistic person. Always see annoying problems behind the problem.

164	Oblatory	奉獻
165	Now act	立即行動
166	With heart	用心
167	Career	生涯
168	sttrive	奮鬪
169	Harmonious	和諧
170	Optimisitic	樂觀
171	Idea	想法
172	Listen	聆聽
173	Excellent	極好的
174	Experience	經驗
175	Crave	勇氣、勇敢
176	Conduct	經營
177	Manage	管理
178	Positive	積極
179	Mental	心理
180	Full	充分
181	Trust	信任
182	Bliss	極快樂
183	expend	費用
184	Expenditure	消費
185	Expert	專家
186	Explain	解釋、說明
187	Expose	曝露

188	Fanciful	空想
189	Farewell	告別，一路平安
190	Fault	缺點
191	Ritter	甘苦
192	Fearful	懼怕
193	Fearfulless	不懼怕
194	Feature	特色
195	Fee	費、報酬
196	Feeble	衰弱
197	Fiction	小說
198	Fierce	兇猛
199	Fight	爭論
200	Figurative	譬喻
201	Firm	商店
202	Flatter	諂媚
203	Float	發行
204	Folly	愚蠢
205	Fluent	流利的
206	Forbear	忍耐
207	Forbidding	形勢險惡的
208	Foremost	首要
209	forgive	原諒
210	Fork	歧路
211	Forsake	遺棄

做人，做事，要做一個有溫度的人，有熱誠的人，有情感的人。
Be a person. Do something. Be a person with temperature. Have a person with enthusiasm.

212	Forth	向前
213	Forward	向前進步
214	Freight	運費貨物
215	Frighten	恐嚇
216	Fulfill	履行
217	Furnish	供給
218	Gay	快活
219	Gamble	賭博
220	Giant	巨人
221	Gleam	光茫
222	Increase	增進
223	Gloom	憂鬱
224	Glory	光榮
225	Young	年青
226	Clothing	衣服
227	Tie	領帶
228	Bed time	就寢時間
229	Lunch	午餐
230	Midnight	午夜
231	Past	過去
232	Cookies	餅乾
233	Fruit	水果
234	Juice	果汁
235	Snake	點心

有關係=沒關係，沒關係，就是要去找關係，找不到關係，就是要去拉關係，拉不到關係，就是有關係。

It does not matter. It does not matter. It does not matter. It is to find a relationship. It does not find a relationship. It is to pull a relationship.

236	Album	相簿
237	Chalk	粉筆
238	Diary	日記本
239	Gift	禮物
240	Sink	水槽
241	Blackboard	黑板
242	Club	社團
243	Game	遊戲
244	Lesson	課
245	Member	組員
246	Race	比賽
247	Subject	科目
248	Hiking	健行
249	Text book	課本
250	Jog	慢跑
251	Sport	運動
252	song	歌曲
253	Harbor	海港
254	Way	通路
255	Fall	秋天
256	Jungle	叢林
257	Seed	種子
258	Adventure	冒險
259	Hero	英雄、勇士

華人獨特的人際關係：

A.認親戚 B.拉關係 C.攀交情 D.做人情 E.鑽營 F.送禮。

The unique interpersonal relationship of the Chinese:

A. Recognize relatives B. Pull relationships C. Cultivate friendship D. Do human relations E. Drill camp F. Give gifts.

260	Celebration	節慶
261	Festiral	假期
262	Holiday	假日
263	Actor	演員
264	Businessman	商人
265	Hunter	獵人
266	Servant	佣人
267	Afraid	害怕
268	Boring	無聊
269	Busy	忙碌
270	Carefully	小心地
271	Cool	涼的
272	Cold	冷的
273	Dangerous	危險
274	Dark	暗
275	Deep	深
276	Delicious	美味
277	Difficult	困難
278	Dirty	骯髒
279	Enough	足夠
280	Expensive	昂貴
281	Far	遠
282	Fast	快
283	Favorite	最愛的

先認識人，再交流，再交心，這是交朋友最佳上策。

Meet people first. Then communicate. Then make friends. This is the best way to make friends.

284	Full	充滿
285	Funny	好笑
286	Glad	樂意
287	Great	很好
288	Advice	勸告
289	Agency	代理店
290	Aim	志向
291	Ally	聯盟
292	Alter	改變
293	Amateurish	業餘
294	Anyhow	任何方法
295	Anyway	無論如何
296	Argament	辯論
297	Ashamed	慚愧
298	Assert	主張
299	Basis	基礎
300	Booklet	小冊
301	Capital	資本
302	Cease	停止
303	Cwaseless	不停止
304	Dancing	跳舞
305	Style	時尚
306	Design	設計
307	Cheat	欺騙

做人，做事，正確比速度更重要。

Being a person. Doing things. Correctness is more important than speed.

308	Chief	主要
309	Commercial	商業
310	Conference	會議
311	Confide	信心
312	Confident	有信心
313	Constancy	有恆心
314	Continance	斷續進行
315	Consign	託運、存
316	Encouragement	鼓勵
317	Encourage	促進
318	Coward	膽怯
319	Creation	創造
320	Credit	信賴
321	Cultivate	栽培
322	Debate	辯論
323	Deal	交易
324	deceit	欺騙
325	Decide	決定
326	Fun decide	未決定
327	Demand	要求
328	Deny	拒絕
329	depend	依賴
330	Describe	敘述
331	Desire	欲望

一個人的成功，15%取決於專業技術，85%取決於人際溝通能力。
One′s success. 15% depends on professional skills. 85% depends on interpersonal communication skills.

332	Determine	決意
333	defermine	決心
334	Determination	專心
335	digestive	消化
336	diligent	勤勉
337	Discuss	討論
338	Distinctive	特色
339	Dovmitory	宿舍
340	Earn	賺得
341	Effort	努力
342	Unending	不停止
343	Enrich	使富足
344	Enterprising	企業心
345	Entry	進入
346	Solve	解決
347	Tycoon	大亨
348	Enforece	執行
349	Enlarge	擴大
350	Equip,emt	設備
351	Enough	足夠
352	Errand	出差
353	Essay	散文
354	Essayist	作家
355	Estate	財產

記錄比億更重要，格局決定決局。

Record is more important than 100 million. The pattern determines the ending.

356	Everything	凡事
357	Everywhere	處處
358	Exclusion	拒絕
359	Excitedly	興奮
360	Exection	實現、執行
361	Exertion	努力、奮發
362	Exhibition	博覽會
363	Exhibit	陳列、展覽
364	Expect	期待
365	Expert	專家
366	Exposition	解釋
367	Failure	失敗
368	Untail	不倦
369	Fade	凋謝
370	Fancy	幻想
371	Fate	命運
372	Fancy	幻想
373	Fault	缺點
374	Faultless	無缺點
375	Fetch	得到
376	Fearful	懼怕
377	Fearfulless	不懼怕
378	fault	過失
379	Fee	報酬

公司運作，決策，效益比效率更重要。
Company operation. Decision-making. Benefit is more important than efficiency.

380	Fight	爭論
381	Final	最後
382	Fire	商店
383	Float	發行
384	Folly	愚蠢
385	Force	強迫
386	Forget	忘記
387	Forture	產業
388	Effort	努力
389	Fortunate	僥倖
390	Frank	直率
391	Freight	運費
392	Friendship	友情
393	Fulfill	履行
394	Further	促進
395	generous	慷慨
396	Glisten	閃耀
397	Grass	青草
398	Gratitude	感激
399	Guarantee	保證人
400	Habit	習慣
401	Hardship	困苦
402	Harmony	和睦
403	Haste	急忙、加速

好心情，好態度，好人緣，就會有好的人際關係。
Good mood. Good attitude. Good relationship. There will be good relationships.

404	Headquarters	大本營
405	Heap	堆
406	Hobby	嗜好
407	Hold	支持
408	Holder	支持人
409	Honest	誠實
410	Horror	深惡痛絕
411	Hospitable	殷勤
412	However	無論如何
413	Humility	謙遜的
414	Healthy	健康
415	Hot	熱的
416	Hungry	餓的
417	Kind	親切的
418	Lazy	懶惰的
419	Less	較少
420	loud	大聲
421	Noisy	吵鬧
422	Polite	有禮
423	Rude	粗魯
424	Silent	寂靜
425	Soft	柔軟
426	Spicy	辣的
427	Thick	薄的

熱愛你的工作，熱愛你的朋友，熱愛你的生活，這是人生成功要素。
Love your work. Love your friends. Love your life. This is the key to success in life.

428	Ugly	醜陋
429	Useful	有用的
430	Wild	粗野
431	Wald	粗野
432	Become	變成
433	Income	收入
434	Borrow	借
435	Burn	燃燒
436	Care	照顧
437	Carry	搬運
438	Change	改變
439	Check	檢查
440	Catch	抓住
441	Choose	選擇
442	Cost	花費
443	Count	計算
444	Fall	跌倒
445	Fix	修理
446	Finish	完成
447	Find	尋找
448	Forget	忘記
449	Guess	猜測
450	Hate	討厭
451	Have	擁有

好的人際關係，會帶來好的人生結局。
Good interpersonal relationship will bring good life ending.

452	Hide	躲藏
453	Hit	擊中
454	Hold	握
455	Hurry	趕快
456	Hurt	傷害
457	Invite	邀請
458	Join	參加
459	Laugh	大笑
460	Listen	聽
461	Miss	錯過
462	Pass	通過
463	Pick	挑選
464	Practice	練習
465	Prepare	準備
466	Pipe	管子
467	Plate	盤子
468	Pleasure	樂趣
469	Point	要點
470	Polite	有禮貌
471	Poor	貧窮
472	Popular	流行的
473	Postcard	明信片
474	Present	禮物
475	Pretty	漂亮

好的溝通能力＋好的人際關係，是事業成功的關鍵所在。

Good communication skills + good interpersonal relationships. It is the key to successful career.

476	Price	價格
477	Prize	獎品
478	Problem	問題
479	Proud	驕傲
480	Route	行程
481	Quick	快
482	Question	問題
483	Quiet	靜
484	Quiz	小考
485	Purple	紫色
486	Ready	準備
487	Record	記錄
488	Repeat	重複
489	Right	正確
490	Road	路
491	Roll	流動、捲
492	Room	房間
493	Rule	規則
494	Sad	悲傷
495	Save	儲備
496	Scared	驚嚇
497	Secretary	祕書
498	Send	寄送
499	Sell	賣

怎麼樣去發掘和開發自己潛在的能量，相信自己，肯定自己，自己是可以的，這是潛能開發最大的推動力。

How to discover and develop your potential energy. Believe in yourself, affirm yourself, and be yourself. This is the biggest driving force for potential development.

500	Serious	認真
501	Share	分享
502	Shine	發光
503	Shop keeper	店主
504	Should	應該
505	Shout	喊叫
506	Shy	羞
507	Sign	簽名
508	Since	自從
509	Sir	老師
510	Sleep	睡覺
511	Smart	聰明
512	Sneek	點心
513	Soon	快
514	Stay	停留
515	Stranger	陌生人
516	Subject	主題
517	Successful	有成就的
518	Sure	確定
519	Surprise	驚訝
520	Sweet	甜
521	Take	拿
522	Taste	嚐試
523	Teach	教學

常常給予真誠的讚美和感謝，這是友誼和人際關係昇華的必備元素。

Always give sincere praise and gratitude. This is an essential element for the sublimation of friendship and interpersonal relationships.

524	Team	隊
525	Television	電視
526	Temple	廟
527	Terrible	很糟
528	Than	比
529	Then	當時
530	These	這些
531	Thing	事情
532	Thirsty	口渴
533	Though	儘管
534	Under	下面
535	Visit	拜訪
536	Wake	醒
537	Warm	暖
538	Without	沒有
539	Word	字
540	Worry	擔心
541	Wrong	錯
542	Answer	答案
543	Yet	可是還沒
544	Able	有本事的人
545	Afraid	害怕
546	gain	得到
547	Angry	生氣

常常引發他人心中的渴望，傾聽他人心中的需求，而給予充分的協助和幫助。

Frequently arouse the desires of others. Listen to the needs of others and give them full assistance and help.

548	Anything	任何事
549	Attack	攻擊
550	Ask	問
551	Bad	壞
552	Smile	笑
553	Begin	開始
554	Believe	相信
555	Boil	沸騰
556	Bow	鞠躬
557	Bright	光明
558	Busy	忙碌
559	Cheer	歡呼
560	Classmate	同學
561	Clerk	售貨員
562	Clear	清楚
563	Deal	交易
564	up	向上
565	Dream	夢
566	During	在期間
567	Early	早
568	Ever	曾經
569	else	要不
570	Excuse	原諒
571	Fan	支持

想法和思維和行動，要確實的落實在平常的生活裡。
deas and thinking and actions. To be surely implemented in ordinary life.

572	Finally	最後
573	Find	找到
574	Follow	跟隨
575	Grown	成長
576	Jobs	工作
577	Lie	謊言
578	Lise	名單
579	Nice	美好
580	Once	一次
581	Pick	挑選
582	Picture	想像
583	over	結束
584	Order	訂購
585	Matter	事情
586	Option	選擇
587	Perhaps	或許
588	Mad	生氣
589	Grasp	把握
590	Seize	攫取
591	Timeliness	天時
592	Land productivity	地利
593	Personal management	人和
594	There fore	所以
595	Official	官方的

小客戶，要當大客戶來服務，來經營，這是業務工作，成功的基本概念。
Small customers. To serve as big customers. To operate. This is the basic concept of business work. Success.

596	Assist	幫助
597	Straw	草
598	Dew	露
599	Struggle	奮鬥
600	Fight	爭取
601	Strive	奮勉
602	Often	時常
603	Frequently	經常地
604	Criticism	檢討
605	Splendid	華麗的
606	Examine	審察
607	Splmdid	極佳
608	Condact	指揮
609	Behave	檢點
610	Affair	工作、事務
611	Mercy	憐憫
612	Benefaction	施惠
613	Philanthropist	慈善家
614	Encourage	鼓勵
615	Impel	催促
616	Urge	激勵
617	Effort	努力
618	Management	管理
619	Hurry	趕快

時時學習，時時增長知識和技能，時時增長智慧。
Learn from time to time. Increase knowledge and skills from time to time. Increase wisdom from time to time.

620	Finish	完成
621	Basic	基本基礎
622	give up	放棄
623	abandon	放棄
624	vigorous	有活力的
625	Enterprising	有進取心的
626	Diligently	勤勉地
627	Cultivation	栽培
628	Accomplish	達成
629	Analysis	分解、分析
630	chant	讚揚
631	Know	認識
632	Familiar	熟悉的
633	Recognize	承認
634	Manpower	人力
635	Wealth	財富
636	Recover	收獲
637	Recapture	收復
638	Harvest	收獲
639	Result	成績
640	Stipulate	規定
641	Fix	規定
642	Provide	規定
643	Law	規律（法律）

愛茵斯坦說，一個人的成功常取決於轉折點上。
Einstein said that one′s success often depends on the turning point.

644	Specification	規格
645	Standard	標準
646	Planning	規劃
647	Discuss	討論
648	Consider	研究
649	Research	研究
650	Study	學習
651	Shortcut	捷徑
652	Norm	準則
653	Punctual	準時
654	Intent	準備
655	Prepare	準備
656	Sense	觀念
657	Idea	觀念
658	Concept	觀念
659	Concerned	關懷
660	Genius	天才
661	Timelines	天時
662	Position	地位
663	Land	地利
664	Productivity	地利
665	Productivity	生產力
666	Personnel	人和
667	Management	人事管理

成功的銷售或服務，就是瞭解並做到超越客戶的期待和期望。

Successful sales or service. It is to understand and exceed customer expectations and expectations.

668	Qentle	和氣
669	Harmomious	和氣
670	Quicken	加快
671	Make a greater effort	加油
672	Immediately	立刻
673	Command	指揮
674	Guide	指導
675	Direction	指導
676	Congratulation	恭喜
677	Positive	積極
678	Active	積極
679	Energetic	積極
680	Rigorous	積極
681	Great	偉大
682	Mighty	偉大
683	Exploit	偉業
684	Dream of	夢寐以求
685	Vainly hope	夢想
686	Philanthropist	善人
687	Mercy	善心
688	Good conduct	善行
689	Kind hearted	善良
690	Good deeds	善事
691	Benefaction	善事

隨時檢視自己的工作效率，想辦法倍增工作效率。
Check your work efficiency at any time. Find ways to increase work efficiency.

692	Good	好的
693	Adepe	內行
694	Rehabititate	精神重建
695	Good will	善意
696	Criticism	善意的批評
697	Project	計劃
698	Excuse	寬恕
699	Forgive	寬恕
700	Foerant	寬恕
701	Plenty	寬裕
702	Luck	幸運
703	Fortunate	幸運
704	Happiness	幸福
705	Felicity	幸福
706	Decide	作主
707	Conduct	作為
708	Accomplish	實現
709	Character	品行
710	Quality	品質
711	Education	教育
712	inculcate	教育
713	Safe	安全
714	Lesson	功課
715	Arrange	整理

銷售之所以成功，就是能掌握市場的需求，大量的行銷概念。

The reason why sales are successful is to be able to grasp the needs of the market. A large number of marketing concepts.

716	Neat	整齊
717	Clean	清潔
718	Hygiene	衛生
719	Sanitation	衛生
720	Promote	提昇
721	Advance	提昇
722	Hoise	提昇
723	elevate	提昇
724	Progress	發展
725	March	進行
726	Enterprising	進取
727	Loyalty	忠心
728	Derotion	忠心
729	Faithfulness	忠心
730	Counsel	忠告
731	Advice	忠告
732	Diligently	用心
733	Conscientious	認真
734	Perfect	完美
735	Consummate	完美
736	Accomplish	完成
737	Overstep	超出
738	Super	超級
739	Cultivation	耕耘

隨時掌握最新的資訊，隨時檢討，分析，做決策，納入公司決策一環，並確實實行之。

Keep up to date with the latest information at any time. Review, analyze, make decisions at any time, and incorporate it into the company′s decision-making process.

740	Life	生命
741	Goal	目標
742	Target	靶子
743	Endure	持久、忍耐
744	Support	鼓勵、支持
745	Courage	勇氣
746	Courtecy	禮貌、謙恭
747	Will	意志力
748	Exceed	超越
749	Efficiency	效率
750	Chance	機會
751	try hard	努力
752	Confidence	信心
753	Knowledge	智識
754	Science	科學
755	Wisdom	智慧
756	Potential	潛能
757	Cater	提供飲食
758	Energy	能量（幹勁）
759	Technic	技術
760	Respect	尊敬
761	Fraternal	友愛
762	Credit	信用
763	Skill	技巧

隨時分析自己的優點和缺點，隨時擴大優點，縮減缺點，使自己變得更優質，更卓越，更精進。

Analyze your own strengths and weaknesses at any time. Expand your strengths at any time. Reduce your weaknesses. Make yourself better, better, more sophisticated.

764	Faith	忠誠
765	Tralatitions	遺傳
766	Honor	榮譽
767	Friendly	友好
768	Friendship	友誼
769	Kind	仁慈
770	Win	勝利
771	Competent	勝任
772	Victory	勝利
773	Triumph	凱旋
774	Eloquence	口才
775	Clever	聰明
776	Intelligent	頭腦靈活口才好
777	Richs	財富
778	Future	未來
779	Prohect	計劃
780	Talent	才能天資
781	Philosophy of life	人生哲學
782	Outlook	前途展望
783	Reward	獎賞
784	Infinife	無限
785	Immeasurable	廣大無限
786	Sharp	犀利
787	Mind	願望

訂立目標，行動計劃，貫徹實行之，終究會成功的，自然的法則，自然的定律。

Set goals. Action plans. Implement them. They will eventually succeed. The laws of nature. The laws of nature.

788	Useful	有用的
789	Service	服務
790	Sale	販賣
791	best choice	最好的選擇
792	Professional	專業
793	Hard	努力、困難
794	To go	加油
795	Goal	目標
796	power	力量
797	Think	想
798	Technic	技術
799	Want	要
800	Develop	開發
801	Place	地方
802	Spread	撒體
803	Friendship	友好
804	Courtesy	禮貌
805	Marketing	市場化
806	Conduct	行為
807	Deliberately	故意的
808	Value	價值
809	Because	因為
810	Understand	了解
811	Gain	利潤、收獲

隨時強化專業技能，隨時強化人際溝通能力，這是成功的養分，也是成功的基石。

Strengthen professional skills at any time. Strengthen interpersonal communication skills at any time. This is the nutrient of success. It is also the cornerstone of success.

812	Always	時常
813	Become	變成
814	Progress	發展、進步
815	Intend	打算
816	Enjoy	享受
817	Reap	收獲
818	Trust	信任
819	Conscientious	有良心的
820	Give	給
821	a lot	很多
822	Guide	引導
823	Direction	方向、指揮
824	Fishing	釣魚
825	Fate	命運
826	Result	結果
827	Game	比賽
828	Advance	退步
829	Advice	忠告
830	Hoist	提升
831	Elevate	提升
832	Despise	輕視
833	Serious	嚴重
834	Tree	樹
835	Bird	鳥

態度＋人際關係是成功的養分，也是成功的保證。
Attitude + interpersonal relationship is the nutrient of success. It is also the guarantee of success.

836	Perch	棲身
837	Rest	休息
838	Recovers	痊癒
839	Harrest	收獲
840	Idea	想法
841	Variety	變化
842	Assets	資產
843	Property	財產
844	Need	需要
845	Effort	努力
846	Opportunity	機會
847	Have	有
848	Foresight	遠瞻未來
849	Goner	失敗者
850	Loser	失敗者
851	Excuse	藉口
852	Find	尋找
853	Method	方法
854	Look	好像
855	Like	好像
856	Sea	大海
857	Lose	失去
858	Ship	大船
859	Hammer	鐵鎚

成功旅途上是艱辛的，是辛苦的，是充實的，是快樂的，絕不退縮，絕不放棄，這是成功唯一的心態和信念。

On the journey of success. It is hard. It is hard. It is full. It is happy. Never back down. Never give up.

860	Elaborate	發揮
861	Yourself	你自己
862	Endure	忍耐
863	Plate	平板
864	Sustain	支持
865	Support	支持
866	God	上帝
867	Fair	公平
868	Everyone	每人
869	Everybody	每人
870	World	世界
871	Only	儘有
872	Nobody	沒人
873	Complete	完全
874	Replace	代替
875	Innovation	革新
876	Process	經過
877	Course	經歷
878	Congratulation	恭喜
879	Reap	收獲
880	Engender	產生
881	Smerge	浮現
882	Strong	強壯
883	Trend	趨向

只要你願意，處處是機會。

As long as you like. Everywhere is opportunity.

884	Step	腳步
885	March	行進
886	Promote	促進
887	Top	頂端
888	Payment	支付
889	Relation	關係
890	Popular	流行
891	Public	公共的
892	Eulogize	稱讚
893	Sweet	甜的
894	Achievement	成就
895	Brain	頭腦
896	Creation	創造
897	Agitate	激蕩
898	Surge	起伏
899	Magnifiene	燦爛
900	Bright	明亮
901	Heaven	天空（神明）
902	Motivation	積極
903	Anything	任何事
904	Aspiration	志氣
905	Aim	輔助
906	Conclude	達成
907	Instant	達成

人因為有理想，生命才有價值。
People have ideals. Life is valuable.

908	Immediately	立刻
909	Best	最好
910	Humble	謙卑
911	Question	問題
912	Enforce	執行
913	Ask	問
914	Operation	操作
915	Organization	組織
916	Root	根本
917	Without	不停
918	Ceaseless	不斷
919	Surmount	突破
920	Break	打破
921	Original	原有
922	accomplishment	成就
923	Hight	高
924	Uncertain	不確定
925	Number	數目
926	Amount	數量
927	Quantity	總量
928	Honest	誠實
929	Faith	忠實
930	Time keeping	守時
931	Keep one's word	守信

成功是靠有準備的，成功是靠有機運的，成功是靠有責任心的，成功是靠有實力的。

Success depends on preparation. Success depends on luck. Success depends on responsibility. Success depends on strength.

932	Rule	規則
933	Regular	有規律
934	Element	要素
935	Make friends	做朋友
936	One's whole life	一生
937	Result	結果
938	Gain	利潤
939	Caution	注意
940	Celerity	迅速
941	Famouscelebrated	著名的
942	Not common	不普通
943	Reply	回答
944	Apartment	公寓
945	Relateding	有關
946	Situation	清洗
947	Headquarters	總部
948	Doctor	博士
949	Professor	教授
950	Socks	襪子
951	Photograph	相片
952	Each	每一
953	Picture	相片
954	Shoes	鞋
955	Vegetables	蔬菜

窮則變，轉彎，變則通。

The poor is the way. The turn. The way is the way.

956	Thicker	厚
957	Underwear	內衣
958	Briefs	內衣
959	Discount	打折
960	10% off	打折
961	Saced	打折
962	Leave	離開
963	On-the-job training	在職訓練
964	Unpaid leave	留職停薪
965	On a business trip	出差
966	On leave	休假
967	annual leave	年假
968	Personal leave	事假
969	Maternity leave	產假
970	Ask for leave	請假
971	Flass	便鞋
972	Slippers	拖鞋
973	Heels	高跟鞋
974	Percent	百分比
975	Building	大樓
976	Industial	工業
977	Extra	額外
978	Juice	果汁
979	Soup	喝湯

走過的路，走過的經驗，就是你的資產。
The road traveled. The experience traveled is your asset.

980	Related	相關的
981	Coffee	咖啡
982	Tea	茶
983	Century	世紀
984	Restaurant	餐館
985	Preschool	幼兒園
986	Partner	夥伴
987	Relate	敘述
988	Silly	憤慨
989	Disappointed	失望
990	Terribly	可怕地
991	Worried	擔憂
992	Naturally	當然的
993	Purpose	目的、效用
994	Elaborate	詳盡的
995	Comment	評論
996	Feedback	回饋
997	Relationship	親屬關係
998	work holism	工作狂
999	Pity	同情
1000	Track	行蹤
1001	Weekend	週末
1002	Trip	旅行
1003	Fault	缺點

肢體語言，影響你的一生，也影響別人的一生。

Body language. Affects your life. Also affects the life of others.

1004	Apology	道歉
1005	Cause	原因
1006	Cruel	殘忍的
1007	congratulations	恭禧
1008	Cheers	乾杯
1009	Christmas	聖誕節
1010	Happiness	幸福
1011	Listen	聽
1012	Expect	意料
1013	Pardon	原諒
1014	Favorite	親信
1015	Change	變化
1016	Variety	多樣化
1017	Convenience	方便
1018	Directly	直接
1019	Personally	當面
1020	Amusement	娛樂
1021	Depend	信賴
1022	Spoil	糟蹋
1023	Appreciate	欣賞
1024	Bother	打攪
1025	Reserbation	保留
1026	Tongye	口才
1027	Engagement	婚約

山不轉，路轉，路不轉，人轉。
Mountain does not turn. Road turns. Road does not turn. People turn.

1028	Mood	心情
1029	Whatever	任何事
1030	Absolutely	絕對的
1031	Actually	實際上
1032	Incorrect	不正確
1033	Quite	完全
1034	Suggestion	建議
1035	Approve	允許
1036	Tough	堅韌的
1037	Unpleasant	不愉快
1038	Message	口信
1039	Transmit	傳達
1040	Comment	評論
1041	Compares	比較
1042	Explain	解釋
1043	obscure	暗的
1044	Reaction	反應
1045	Elaborate	精心製作
1046	Mess	伙食團
1047	Disappointed	失望
1048	Frightened	受驚
1049	Exactly	確切的
1050	Indeed	對的
1051	Mention	提到

時時面對人，事，物，隨時要保持感謝的心，感恩的心。

Face people, things, things. Always be thankful. Be grateful.

1052	Lately	最近
1053	Roommate	室友
1054	Probably	可能
1055	Exhausted	力量
1056	Famous	著名的
1057	Allow	允許
1058	Pack	裝入
1059	Lie	謊話
1060	Baggage	隨身行李
1061	Serve	供應
1062	Plain	清淡的
1063	Ceremony	禮儀
1064	Prefer	較喜歡
1065	Jungle	叢林
1066	Fortune	命運
1067	Fond	喜歡
1068	Recently	最近
1069	Common	普遍
1070	Cart	推車
1071	Observe	看到
1072	Comfortably	舒適的
1073	Consequently	因此
1074	Poison	毒
1075	Discourage	氣餒

肯定+讚賞是自信的培養土，休息是為了走更遠的路。

Affirmation + appreciation is the cultivation ground of self-confidence. Rest is to go a long way.

1076	Shoplift	順手牽羊
1077	Earn	賺
1078	Pass on to	轉移到
1079	Pay for	付款
1080	Pay off	還清
1081	Realize	實現
1082	Scientist	科學家
1083	Conclude	斷定
1084	Look after	照顧
1085	Take care of	照顧
1086	Refer to	提交
1087	Come true	實現
1088	A school of	一群
1089	Translation	譯文
1090	Explanations	說明
1091	Host	房東
1092	Hospitality	熱情接待
1093	Reading	標題
1094	Salutation	稱呼
1095	Signature	簽名
1096	Representative	代表
1097	Relative	親戚
1098	Robot	機械人
1099	Classes	上課

失敗，挫折，打擊，羞辱，都是你再一次成長的養分。
Failure, frustration, thrashing, humiliation are all the nutrients you grow up again.

1100	on vacation	在度假
1101	Recipient	收信者
1102	Suddenly	突然地
1103	Practically	幾乎
1104	Permission	許可
1105	Atmosphere	氣氛
1106	Irritate	刺激
1107	Go mad	發瘋
1108	Impression	印象
1109	Preparation	準備
1110	Ruin	毀掉
1111	Salary	薪水
1112	Expect	預期
1113	Plenty	許多
1114	Regulations	規定
1115	Torch	火炬
1116	Celebrate	慶祝
1117	Statue	雕像
1118	Athlete	運動員
1119	Normally	正常地
1120	Exist	存在
1121	Recent	最近
1122	Tournament	比賽
1123	Trap	陷阱

工作的開始，正是學習和磨練的開始，隨時保持成長進步和超越。

The beginning of work. It is the beginning of learning and discipline. Keep growing and progressing at any time.

1124	Path	通路
1125	Aspect	方面
1126	Affect	影響
1127	Humanity	人類
1128	Dispensable	不必要
1129	Imagine	想像
1130	To sum up	總之
1131	Ambition	野心
1132	give up	放棄
1133	In vain	徒然的
1134	Memorize	記憶
1135	Persevering	堅忍
1136	Tend	走向
1137	Freelu	隨意的
1138	Fluent	流利
1139	Academic	學識
1140	Sit up late	熬夜
1141	Boredom	厭倦
1142	Inexpensive	所費不多
1143	Differensive	差異
1144	Reduce	減少
1145	Modify	改變
1146	Diet	飲食
1147	Adopt	採用

校長，企業家，領導人，就像花圃的園丁，天天灌溉每一顆種籽，天天照顧每一顆種籽（員工），快快成長，健康壯大。

Principal. Entrepreneur. Leader. Just like the gardener of the flowerbed. Irrigating every seed every day. Taking care of every seed every day （employee）. Growing fast. Healthy and strong.

1148	Intend	打算
1149	Require	需要
1150	Thoughtful	體貼
1151	Financial	財務
1152	Aid	援助
1153	Make pulice talks	做演講
1154	Contact	接觸
1155	Note	注意
1156	Spoil	破壞
1157	Appreciation	欣賞
1158	Vary	變化
1159	Set forth	發表
1160	Acquaint	熟悉
1161	Exact	正確
1162	State	起源
1163	Connection	關係
1164	Concentrate	專心
1165	Consequence	結果
1166	Frequent	頻繁
1167	Phenomenon	現象
1168	Prosperity	繁榮
1169	Interflow	交流
1170	Source	來源
1171	Preserve	保護

只要肯上進，肯努力，貧窮也能變富有。
As long as you are willing to make progress, be willing to work hard, and poverty can become rich.

1172	Band	隊
1173	Donate	捐款
1174	Virgin	未開墾
1175	Taise	籌款
1176	Plant	廠
1177	Benefit	好處
1178	Definite	明確
1179	Steady	固定
1180	Pace	步伐
1181	Stress	強調
1182	after all	畢竟
1183	Reckless	魯莽
1184	Severe	嚴厲
1185	Punishment	懲罰
1186	Modest	謙虛
1187	Despite	即使
1188	Exceptional	例外
1189	Count on	信賴
1190	Purpose	目的
1191	effort	努力
1192	action	行動
1193	attitude	態度
1194	actionist	行動主義者
1195	alwaysg	時常

只要肯用心，肯下功夫，肯學習，處處是機會。

As long as you work hard. You work hard. You learn. Everywhere is an opportunity.

1196	anything	任何事物
1197	able	能夠的
1198	accomplish	完成
1199	abandon	放棄
1200	anytime	任何時間
1201	advance	進步
1202	a lot	大量
1203	assets	資產
1204	alter	改變
1205	ambition	雄心
1206	achievement	成就
1207	agitate	鼓動
1208	asplration	渴望
1209	aim	目標
1210	ask	問
1211	accomplishment	成就
1212	Best	最好的
1213	business	生意
1214	because	因為
1215	Become	變成
1216	blossom	繁榮
1217	basic	基本的
1218	believe	相信
1219	bird	鳥

只要你願意，成功大門，永遠為你而開。

As long as you are willing. The door to success. Always open for you.

1220	beautifully	美麗地
1221	bitter	苦味的
1222	brain	頭腦
1223	bright	明亮的
1224	Byeak	破裂
1225	choice	選擇
1226	chance	機會
1227	courag	勇氣
1228	Confidence	信心
1229	Curtesy	禮貌
1230	conduct	行為
1231	Ceremery	禮儀
1232	Come	來
1233	Createion	創造
1234	Cspital	首都、資本
1235	Consummate	圓滿的
1236	Cultivation	栽培
1237	Consider	考慮
1238	Complete	終止、完整的
1239	Course	過程、進行
1240	Congratulation	祝賀
1241	Conscientious	認真的
1242	compare	比較
1243	Change	改變

只要我願意，我的未來前景，前途是很有希望的。
As long as I want. My future prospects. The future is very promising.

1244	topic	題目、主題
1245	Constancy	忠誠
1246	Continuance	繼續
1247	Conclude	結束、訂立
1248	correct	正確的
1249	Ceaseless	永不停的、無限的
1250	Develop	發展
1251	Deliberately	慎重地
1252	Decide	決定
1253	Don't give up	不放棄
1254	Difficult	困難的
1255	Discuss	討論
1256	Definite	一定的
1257	Different	不同的
1258	Dispensser	藥劑師、分配者
1259	Drink	唱
1260	Design	設計
1261	Diligently	勤勉地
1262	Diligent	勤奮地
1263	Direction	方向、指揮
1264	Determine	裁定
1265	Despise	輕視
1266	Dream	夢
1267	Degree	程度

只要你願意，泥土也會變黃金。

As long as you want, the soil will become gold.

1268	Detail	詳細
1269	Destiny	命運
1270	Immediate	立刻
1271	Energy	精力、能量
1272	Effectiveney	效率
1273	Everywhere	到處
1274	enjoy	享樂
1275	Elevate	抬高、提高地位
1276	Effort	努力
1277	Exclusive	除外的
1278	Explarge	擴大
1279	Enough	足夠的
1280	Enterprising	有事業心的
1281	Excuses	藉口、原諒
1282	Elaborate	精心製作的
1283	endure	容忍、持久
1284	Faith	信用
1285	Friendly	友善的
1286	Friendship	友誼的
1287	Future	未來
1288	Finish	完成
1289	Familiar	熟悉的
1290	First	第一
1291	Further	進一步的

常常懷著，讚美，感謝，感恩的心。
Often with. Praise. Thanks. Grateful heart.

1292	Foot	腳
1293	Fetch	拿來
1294	Fee	費用、賞金
1295	Filteration	過濾
1296	Find	發現
1297	Fair	公平的
1298	Fishing	釣魚
1299	Fate	命運
1300	Foresight	遠瞻未來
1301	friend	朋友
1302	Goal	目標
1303	Gain	得到、利潤
1304	Gentle	溫柔的
1305	Goner	失敗者
1306	God	上帝
1307	Give	給
1308	Guide	導引
1309	Games	遊戲、比賽
1310	Glisten	輝耀
1311	Grow	生長
1312	Health	健康
1313	Happy	快樂
1314	Hard work	勞力工作
1315	Hurry	趕快

1316	Honest	誠實
1317	Hoist	舉起、起重機
1318	Harmony	融洽
1319	Hammer	鐵鎚
1320	Hereafter	將來
1321	Harvest	收獲
1322	Have	有
1323	hope	希望
1324	Hike	徒步
1325	Heaven	天空、神
1326	Help	幫助
1327	However	無論如何
1328	humble	謙遜的
1329	Finite	無窮的
1330	Inceme	定期收入
1331	Intend	打算
1332	Idea	主意、理想
1333	If	假使
1334	Impartial	公平的
1335	Innovation	創新
1336	Idly	懶惰地
1337	Instant	速成的
1338	Immediately	立刻
1339	Know	知道

走出舒適圈，突破舒適圈，改變舒適圈。

Out of the comfort zone. Break through the comfort zone. Change the comfort zone.

1340	Kind	仁慈的、種類
1341	knowledge	知識
1342	Life	生活
1343	Lifeblood	鮮血、力量的泉源
1344	Lazy	懶惰的
1345	Leaning	傾向
1346	Lose	失去
1347	Lead	領導
1348	Loser	失敗者
1349	Look	看
1350	Like	喜歡
1351	lichen	青苔
1352	Money	錢
1353	Market	市場
1354	Method	方法
1355	Manpower	人力
1356	Matter	物質、事情
1357	Merit	長處
1358	March	前進
1359	Medium	中間、手段
1360	Myself	我自己
1361	Magnificent	壯麗的
1362	Motibation	刺激、推動
1363	Moss	苔蘚

1364	mentally	心理上地、用腦力地
1365	Nobody	無人
1366	Need	需要
1367	Now	現在
1368	never	永不
1369	Outlook	前途
1370	Opportunity	機會
1371	Official	官方的、正式的
1372	Opinion	意見
1373	Overstep	超出
1374	Only	只有、唯一的
1375	Other	其他的
1376	Operation	操作、手術
1377	organization	組織
1378	Power	權力、動力
1379	Price	價格
1380	Project	計劃
1381	Place	地方
1382	People	人們
1383	Positive	正面的、積極的
1384	Progress	發展、前進
1385	Plan	計劃
1386	Promote	促進
1387	Point	點

傾聽+讚美+重視+尊重+敬佩=好的人際關係
Listen + praise + value + respect + admiration = good interpersonal relationship

1388	Professional	專業的
1389	Pure	純正的
1390	Person	人
1391	Perform	實行
1392	Promote	推廣
1393	Pretext	藉口
1394	Plate	盤、金屬板
1395	Perfect	完美的
1396	Process	過程、程序
1397	Possible	可能的
1398	Perch	棲息
1399	Property	財產
1400	Position	位置
1401	Present	現在的、出席的
1402	Proceed	進行、繼續
1403	Payment	付款
1404	Popular	流行的、一般的
1405	Public	公共的
1406	Psychology	心理學
1407	Physically	身體上的
1408	Peace	和平
1409	Quicken	鼓舞
1410	Quiet	安靜的
1411	Question	問題

合諧就是力量，合作就是力量，團隊就是力量。
Harmony is power. Cooperation is power. Team is power.

1412	Rich	富有的
1413	Respect	尊敬
1414	Right	右的、正確的
1415	Road	路
1416	Recognize	認可
1417	Resolve	堅定
1418	Replace	代替
1419	Renovation	革新
1420	Result	結果
1421	Race	競賽
1422	Rest	休息
1423	Recovery	痊癒、回收
1424	Reap	獲得
1425	Ready	準備好的
1426	Relation	關係、親戚
1427	Root	根
1428	rolling	轉動的
1429	Speed	速度
1430	Success	成功
1431	Science	科學
1432	Sale	出售
1433	Spread	塗布、展開
1434	Service	服務
1435	Schedule	目錄、計劃表

要訓練好笑容，要訓練好相處，要訓練好合群。

To train a good smile. To train to get along well. To train a group.

1436	Something	某事
1437	Skill	熟練
1438	Speeder	加速器
1439	Step	腳步
1440	Serious	嚴重
1441	Success	成功
1442	Study	學習
1443	Sense	感覺
1444	Sea	海
1445	Ship	船、艦
1446	Silent	安靜的
1447	Sustain	撫養、忍耐
1448	Support	支持、鼓勵
1449	Successful	成功的
1450	Standpoint	立場
1451	Scorn	輕視
1452	Strong	強壯的
1453	Sweet	甜的
1454	Surge	大浪
1455	Stop	停止
1456	Surmount	戰勝、超越
1457	Stone	石頭
1458	society	社會
1459	Try hard	試著努力

人脈是致富的跳板，人脈不一定是錢脈，而是要有正確的人脈，才正是錢脈。

The network is a springboard for getting rich. The network is not necessarily the money. It is the right network. It is the money.

1460	To go	去進行
1461	trchnic	技術
1462	Time	時間
1463	Targer	目標
1464	Trust	信賴
1465	Teach	教導
1466	Tree	樹
1467	Trend	趨勢
1468	Triumph	勝利
1469	top	頂端
1470	Inderstand	了解
1471	Unlimited	無限的
1472	undertaking	保證
1473	Value	價值
1474	Vigorous	有活力的
1475	Vision	視力、眼光
1476	Vitality	失命力
1477	variety	變化
1478	Will	將、意志
1479	Want	要
1480	Wealth	財富
1481	Way	道路
1482	World	世界
1483	Wish	願望

人脈是致勝的關鍵，也是事業成功的保證。

Networking is the key to winning. It is also the guarantee of career success.

1484	Willing	樂意的
1485	Without	除外
1486	Yes, I do	是，我願意
1487	Yourself	你自己
1488	Yes, I will	是，我將會

NOTE

NOTE

國家圖書館出版品預行編目資料

超級銷售致富祕訣 Secret of getting rich through super sales／徐錦坤編著. --初版.--
臺中市：徐錦坤，2020.11
面；　公分
中英對照
ISBN 978-957-43-8188-3（平裝）
1.銷售 2.職場成功法
496.5　　109015972

超級銷售致富祕訣

Secret of getting rich through super sales

編　　著　徐錦坤
校　　對　徐錦坤
出版發行　徐錦坤
E-mail：andy0932504612@gmail.com
購書：劃撥帳號：02738961　　戶名：徐錦坤
台中地區農會帳號：62509110020721
訂購電話：0932504612　　※寄送或宅配免運費
設計編印　白象文化事業有限公司
專案主編：黃麗穎　　經紀人：徐錦淳
經銷代理　白象文化事業有限公司
412台中市大里區科技路1號8樓之2（台中軟體園區）
出版專線：（04）2496-5995　　傳真：（04）2496-9901
401台中市東區和平街228巷44號（經銷部）
購書專線：（04）2220-8589　　傳真：（04）2220-8505
印　　刷　唐采印刷事業股份有限公司
初版一刷　2020年11月
優 惠 價　600元

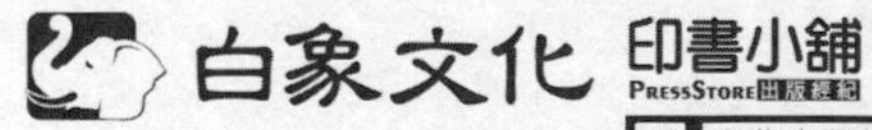